ACTION RESEARCH FOR SUSTAINABILITY

Shared meeting in the Mayville Community Gardens

Source: Eames et al. 2009.

Reprinted from the Citizen Science for Sustainability project.

In gratitude to my family

Action Research for Sustainability

Social Imagination Between Citizens and Scientists

JONAS EGMOSE
Roskilde University, Denmark

Routledge
Taylor & Francis Group
LONDON AND NEW YORK

First published 2015 by Ashgate Publishing

2 Park Square, Milton Park, Abingdon, Oxfordshire OX14 4RN
52 Vanderbilt Avenue, New York, NY 10017

Routledge is an imprint of the Taylor & Francis Group, an informa business

First issued in paperback 2019

British Library Cataloguing in Publication Data
A catalogue record for this book is available from the British Library.

The Library of Congress has cataloged the printed edition as follows:
Egmose, Jonas.
Action research for sustainability : social imagination between citizens and scientists / by Jonas Egmose.
pages cm
Includes bibliographical references and index.
ISBN 978-1-4724-4282-6 (hardback)
1. Environmental sociology. 2. Sustainability--Social aspects. 3. Action research.
4. Environmentalism--Social aspects. 5. Science--Social aspects. I. Title.
GE195.E34 2015
333.72--dc23

2014030044

ISBN 978-1-4724-4282-6 (hbk)
ISBN 978-0-367-33288-4 (pbk)

Contents

List of Figures

About the Author

Jonas Egmose works as an action researcher with a strong research interest for new forms of knowledge creation in the cross-field between sustainability and democracy. He holds the position as Assistant Professor at the Department of Environmental, Social and Spatial Change at Roskilde University, Denmark, where his teaching in particular concerns citizens' participation and urban sustainability. Being part of the Danish Centre for Action Research and Democratic Societal Change his research concerns how democratisation of knowledge creation can help enabling sustainable ways of living. Formerly he has worked as a researcher at the Citizen Science for Sustainability project at Brunel University, London, and in the field of cross-European citizen participation at the Danish Board of Technology. Jonas holds a PhD in Social Science, a Master in Technological and Socio-Economic Planning and a Bachelor in Philosophy and Natural Science Studies from Roskilde University.

Foreword

Mary Brydon-Miller
University of Cincinnati

In Jonas Egmose's view the response to the crucial need to address issues of sustainability lies in the ability of scientists and citizens to engage in democratic processes to develop shared understandings and responses that allow us to fundamentally rethink the way in which we live together in the world. The notion of upstream public engagement, that is the idea that public engagement should take place not simply in response to scientific research but should be brought to bear in framing the kinds of questions that scientists are asked to engage on our behalf, is a critical reframing of this relationship.

In developing this more engaged and generative response to the issue of sustainability we must step outside of the normal strategies of doing science and our accepted ways of relating to one another and to the wider world. This sounds like an overwhelming challenge and to some extent current political inaction on both local and global scales would suggest that the most likely response is continued inertia. But at the same time there is growing understanding in both the scientific and broader community discourses that continued inaction will only exacerbate the environmental, social, economic and political impacts we will face in the future. But who decides what sustainability looks like and what responses will do the most to enhance the flourishing of all life on our Earth?

This volume makes important contributions to this discussion on multiple levels. At the most concrete level, Egmose provides an engaging description of an innovation attempt to bring diverse groups of community members from a low-income neighbourhood in Islington in North London together with academic researchers interested in exploring sustainability at the local level. Working with groups of youths, older people and women aged 25–40 years of age the research sought to use film-making and other forms of community-engaged research to address the question, 'what it is like to live in the local area?', using this question to encourage participants to begin to envision together what a sustainable future might look like in their community. The answers as explored in the films each group created and the discussions which grew out of these collaborations present the range of experiences and aspirations of the participants. For the women's group the main theme seems to be the importance of relationships with one another; for younger people, the feeling of being trapped by the economic and social constraints they face in their daily lives; and for the elderly it was a more aspirational tone characterised by the words, 'What a wonderful world it would be'. (I don't know if it's because I am closer to their age that when I hear this I think of Louis Armstrong's wonderful song with that title.) These distinct responses to the initial question all challenged the participating scientists to rethink the ways in which the notion of sustainability manifests itself in the lives of these community members.

In addition to this finely crafted examination of practice, Egmose also provides an excellent example of the ways in which theoretical frameworks can be used to better understand the nature of human interaction and social change. Building upon the metaphor of painting, Egmose notes that 'the choice of theoretical frameworks is a process of deciding what layers and perspectives are relevant for examining the questions in mind'. By integrating aspects of the work of environmental theorist Vandana Shiva, together with critical theory and Critical Utopian Action Research, Egmose provides a nuanced integration of theoretical perspectives through which to understand the dynamics of the Citizen Science for Sustainability (SuScit) project and its outcomes. Shiva's work contrasting the notion of intellectual commons and enclosures provides an especially useful critique of the current system of knowledge production and the ways in which this constrains the development of creative and socially responsive strategies to address the issue of sustainability.

Perhaps the most radical aspect of this book, however, is in Egmose's willingness to challenge the entire system of knowledge production which currently holds sway within the academy. One important outcome of the SuScit process as Egmose describes it is in the opportunity it affords not just the citizens involved, but the scientists as well, to experience a kind of 'free space' in which the quotidian duties and organizational expectations of academic research are set aside for a time, allowing participants to consider together what a more sustainable future might look like, or as Egmose puts it, to think 'beyond the horizon of the present'. This free space challenges the system of enclosures described by Shiva and the pernicious system of knowledge commodification that currently shapes the way in which science is conducted.

In my mind the key challenge raised by Egmose's work is for us to find ways to create permanent free spaces in which community members and their academic partners might engage with these critical issues together. As Egmose notes, the challenges facing us are not something that scientists can solve through the imposition of a series of technical fixes. Rather we must understand that 'sustainability in its most basic sense is an immanent and emergent ability of ecological and social life continuously to renew itself. No technology can guarantee that ability, but technology can, as any other human invention, either support or erode that basic ability'. Egmose provides here a well-supported and richly illustrated examination of this critical question and offers us both theory and practice designed to help us to tackle these issues and, in doing so, to redefine the role of scientists in our societies as partners in creating positive social and environmental change.

Mary Brydon-Miller, PhD, directs the University of Cincinnati's Action Research Center and is Professor of Educational and Community-based Action Research in the Educational Studies Program in the College of Education, Criminal Justice and Human Services. She is a participatory action researcher who conducts work in both school and community settings. She recently completed work on the *SAGE Encyclopedia of Action Research* with co-editor David Coghlan. Other recent publications focus on the development of new frameworks for understanding research ethics in community settings, including chapters in the *Handbook of Social Research Ethics* and the *Handbook of Action Research.* She is a member of the editorial board of *Action Research* and has co-edited Special Issues of the journal on *Ethics and Action Research and Arts-Based Action Research.* She is also an active contributor to the new social media site, Action Research Plus, which is developing internet-based strategies for disseminating research and building networks. She recently completed a Fulbright Research Fellowship at Keele University in the United Kingdom where her work focused on developing new strategies to inform the ethical conduct of community-based research.

Preface

This book has emerged from numerous conversations sharing the theme that furthering sustainable ways of living is not least a democratic challenge. It would never have been written without the insights shared among each of the citizens, scientists, practitioners and organisers being part of creating the Citizen Science for Sustainability (SuScit) project, which provides the empirical framework for this book. In relation to developing the analytical framework I deeply thank my good colleagues at Roskilde University and at our Danish Centre for Action Research and Democratic Societal Change working in the tradition of Critical Utopian Action Research for the support and inspiration you have brought to this work. Finally, my thoughts are with the memory of Kurt Aagaard Nielsen, being a true mentor over the years: his ways of addressing the basic democratic question – *How do we want to live?* – continues to inspire for furthering democratic culture.

Jonas Egmose, February 2015

Chapter 1
Introduction

Furthering sustainability perhaps represents one of the oldest and yet most complex challenges faced in human history: to enable ways of living, which can be sustained on Planet Earth. This book concerns how action research might have a role to play enabling new forms of social learning and imagination between citizens and scientists in order to address sustainability as a democratic challenge.

The challenges of sustainability confronting modern societies do not only emerge from the inherently interrelated nature of societal and environmental dynamics. They involve a true paradox of modernity: that scientific thought is not only offering solutions to, but evidently making up an inherent part of the socio-technological dynamics constituting present states of unsustainability. The aim of this book is twofold: to reflect on what it might imply for the nature of science to aim at a type of development which is equally socially and ecologically sustainable; and to experiment, concerning how action research might help enabling such modes of science come into being.

Science and Sustainability

The term sustainability is often criticised for being used, and misused, as an empty concept covering almost everything from society to the environment. Whilst it holds true that common use of the term often appears rather vague, it is equally true that the concept implies a strong potential in conceptualising social and environmental problems as highly interrelated challenges. In my perspective the particular strength of the concept of sustainability is that it provides a language for addressing what I understand as *social environmental problems*: historical challenges of our time, which must be understood as equally socially and environmentally grounded.

The nature of such problems has often been described as multidimensional (Beck 1992): not only do they exceed the barriers between local, regional and global scales, and transcend intellectual divides (and scientific disciplines) between explaining the laws of nature and understanding societal dynamics, they are also highly correlated to the socio-technological dynamics which follow the development of the modern society.

Historically, the development of modern society is often described as being closely interlinked with the historical emergence of Science, referring to scientific traditions of systematised exploration and rational argumentation advancing from the so-called Scientific revolution of the seventeenth century. However, it is worth noting that science was not conceptualised as 'Science' (with a capital S) until the *industrial* revolution in the mid-nineteenth century[1] (Postman 1992: 176). Accordingly, one may ask what role Scientific thinking might play for moving towards sustainability, taking into account the historical evidence that Science not merely offers solutions to, but simultaneously makes up an inherent part of the modern socio-technological dynamics from which social environmental problems seem to emerge. Historically scientists, philosophers, sociologists, politicians and citizens have continuously discussed this paradox in various forms through deliberations on the role of Science in society. A number of interesting attributes to Science can be highlighted in this particular context. Taking into account the expanding number of scientific disciplines it is obviously difficult to ascribe *a* common set of characteristics to 'Science' in general. However, insights from philosophy and sociology of science at least suggest some prevalent features of the societal role of Science. Grounding the framework of critical theory Adorno and Horkheimer (1972) in particular addressed the philosophical question, what Science has to offer as a societal strategy and what problems it implies. Later, and perhaps at a more concrete level, Beck (1992) in particular has been addressing relations between Science and the environmental crisis as flipsides of scientific progress. In these and similar societal conceptualisations a number of key features are often ascribed to Science. Firstly, an often mentioned characteristic of Scientific explanation is the underlying aim to produce non-context-specific universal knowledge (and much Scientific methodology might be seen as the attempt to develop paradigmatically consistent criteria for doing so). Secondly, Science has historically developed into ever more specialised sub-disciplines providing increasingly detailed descriptions by the division into still more narrow research fields. A third common characteristic of Science is the underlying strategy of 'mastery of nature': the ability to explore causal relations in order to manipulate material processes. Fourthly, which might appear somewhat contradictory to the above, Scientific communities often strongly argue for a strict distinction between core and applied Science; between the process of Scientific research and the societal application of its outcomes. Although often taken for granted,

1 According to Postman (1992) the term Science and the distinction between 'scientific knowledge' and 'knowledge' first appears in language in the period of the *industrial* revolution. As noted by Nielsen (1996) the introduction of the word Science also reflects the emergence of expert-culture and the role of experts as constituting for industrial development.

this distinction, however, is far from unproblematic. Rather it has become increasingly clear that neither scientific knowledge nor technologies can merely be seen as innocent entities, but must be understood as societal artefacts, framed and constituted in particular historical settings; inherently reflecting certain values and aims; and holding certain often unpredictable potentials.

Keeping this in mind, how do the historically constituted attributes of Science outlined above correspond with the known dynamics of social environmental problems? In my perspective Science without any doubt has a central role to play in describing environmental dynamics thereby enabling the quest for possible solutions to these challenges. However, it is becoming still clearer that any such solution *will* include societal changes. There is simply no way in which we can talk about moving towards sustainability without taking into account the interrelated dynamics between environmental and social change. This is not new to academics. For decades cross-fields between environmental and social sciences have been analysing changes needed, developing technical solutions and maturing strategies to move forward at societal level (Manzi et al. 2010). Today's paradox is not a lack of well-described possible pathways or technical opportunities to deal with environmental challenges; simply, it is the still unanswered question *how to make that happen* in modern society. Among academics, the concept of enabling behavioural change is often used in this context. On the one hand I find this notion useful because it underscores that cultural change is needed. On the other hand I find it insufficient if it implies that this should happen primarily on the basis of rather deterministic approaches of technocratic steering mechanisms responding to environmental needs. Because social environmental problems basically concern the way modern life is organised, it is essential to discuss sustainability not least as a cultural issue, and a challenge to modern democracies. It is in this particular context action research might have a role to play.

The outset for this book is that it does not make sense to discuss environmental or societal change without talking about social change. Without taking into account the dynamics of social change we are not going to enable the cultural and behavioural change to modern consumption, which is key to the environmental discussion. Social dynamics, however, are not universal. They are spatially and temporally grounded in historical, cultural contexts, which need to be taken into account. Hence, to make social change happen, local approaches are necessary for global adaptation.

This particular aspect of moving towards sustainability calls for new approaches by which the scientific and the social can mutually take part in enabling societal change. Clearly there are discrepancies between, on the one hand, the above outlined Scientific norms of non-contextual universal knowledge, disciplinary thinking, the ethos of mastery of nature, the clear-cut distinction between Science and its applications; and on the other, the complexity of the societal encounter with the environmental challenges transcending the local and the global, the call for inter- or even trans-disciplinary thinking, and the historical fact that Science and its applications have been playing a double-edged role continuously, both taking part in creating and solving the problems we are facing today. Hence, the historical challenges of sustainability clearly call for examining new modes of interaction between science and society.

In order to discuss action research in this context, this book is written within the particular research field of studying relationships between science and society; an area historically emerging from and made essential by the manifold historically constituted cultural, environmental and societal challenges of science, some of which have been outlined above. Compared to many contemporary studies in this field (such as STS or Science Technology Studies), it is essential to me that these issues cannot be reduced to questions primarily about science and technology. Most basically they are *societal questions* because they concern how we live, and how we want to live, and therefore questions which in a modern democratic society call for democratic deliberation.

Public Engagement in Science

One locus for democratic deliberations on the relations between science and society is initiatives on public engagement in Science. Whilst it is obviously far from the only interesting point, it is, at least in a European context (European Commission 2001; 2002; 2013) an increasingly important one, which has in recent decades fostered a rising number of initiatives and pilot projects democratically experimenting with new forms of dialogues across citizens and scientist. Consequently the field of public engagement has developed from primarily being an arena for communicating science to the public, towards an experimental site for the interrelation between science and society. Although it is difficult to measure the impact of such public engagement initiatives accurately, the contribution so far seems to be twofold. On the one hand, public engagement initiatives can be seen as part of what have been described as a democratic development, where Science has no longer a privileged right to refer to paradigmatic self-referential criteria when arguing for the value of science in a societal context (Gibbons et al. 1994). On the other hand, public engagement initiatives that might at first appear to broaden discussions at discursive micro levels still tend to reproduce predefined discourses of privileged power on the macro level (Cooke and Kothari 2001). For instance, it has been argued that public engagement initiatives often reproduce an expert-based risk-discourse rather than honouring the real potentials of societal deliberation on Science and technology by adding new dimensions to the discussion (Wynne 2006). Adding that the starting point for most public engagement initiatives is the scientific or technological development itself, not the societal context and challenges in which Science is only one part, public engagement initiatives often seem to be designed in ways in which Science will always be one step ahead of the societal discussion, which therefore can only ad hoc add new layers to predefined technological pathways. The implication is that public engagement effectively runs the risk of legitimising science rather than addressing the fundamental democratic question of how we want to live (Nielsen and Nielsen 2006b).

Despite these teething troubles these trends of public engagement are in a broader historical perspective by no way extraneous. In times where we do not have any final answers to the social environmental challenges we are facing, the experimental approach of public engagement in particular offers opportunities for exploring new and different ways to deliberating on the role of science in society. No public engagement initiative so far has claimed to deliver the one and only right way to organise such new types of deliberations. On the contrary the practice of public engagement is a field occupied by numerous experimental tools, methods, approaches and best practice. Although practitioners of public engagement often argue for the successfulness of their own work, the real long-term potential might lie within the current manifold and experimental nature of this field. When I highlight this aspect of public engagement in particular, it is in order to address another crucial starting point for this book. That for future public engagement to cope with the above outlined complexities it is insufficient merely to develop new social technologies by which we try to force through democratic legitimacy for technological pathways. The real potential and challenge of public engagement is to work and create changes *with* people, not as research objects but as citizens in a democratic society. To experiment, not only about things we already know, but also about ideas that might not yet have become fully conceptualised, and strive to bring these more openly in dialogue with diverse perspectives on what is perceived as desirable futures, as important aspects of the ambiguous societal contexts in which contemporary science emerges. In this perception public engagement exactly means *public* engagement, which can per se only take place through a democratic sphere.

As already emphasised it is evident from numerous studies that public engagement and real involvement is far from the same. Accordingly the academic challenge furthering involvement in public engagement is not merely to describe what is already happening (or not happening) in various public engagement set-ups, but crucially to experiment in order to explore modes of involvement, which we haven't already seen. To do so, in ways that are academically consistent, implies the use of scientific methodologies, which are adequate for experimenting with and exploring the social dynamics of such initiatives. While social science offers innumerable methods for exploring, understanding and describing social dynamics, experimenting with involvement in particular calls for the methodology of action research, which has been the main approach of this work. Doing action research basically is a question of working *with* people aiming to create social change. Action research transcends usual distinctions between the researcher and the researched and invites participants ideally to take part in a mutual process of learning and change, which might enable new modes of thinking and acting. Numerous traditions of action research today offer consistent methodologies used around the world (Reason and Bradbury 2008; Greenwood and Levin 2007; Nielsen and Svensson 2006). Clearly the methodological approaches of action research are rather different from the knowledge paradigms of Science, and from many other disciplines studying public engagement. In many ways action research transcends mainstream scientific approaches. Doing so obviously requires a high level of reflexivity and consideration in order to be academically consistent. What action research offers are methodologies whereby knowledge is not just produced by academics within the universities; rather knowledge creation is something happening among and with all those taking part in the process. Hence, if contemporary social environmental problems call for new ways to approach the relation between science and society, action research offers alternative approaches to knowledge creation, which are both visionary and critical.

Towards New Modes of Science?

In recent years conceptualisations of Science and public engagement have been confronted by new understandings theorised by various scholars in slightly different ways: the move from science towards research (Latour 1998); the development from normal science towards post-normal science (Funtowicz and Ravetz 1993; 1999); the concept of the triple-helix collaboration between academia, industry and the state (Shinn 2002). Although these different conceptualisations are continuously used and further developed, one theory in particular has had an impact on the academic discussion of public engagement: the idea of an historical development from mode-1 towards mode-2 science, often referred to as *a new mode of knowledge production* (Gibbons et al. 1994; Nowotny, Gibbons and Scott 2001). Whilst mode-1 science is constituted by university scientists using specific academic disciplines for developing new theoretical knowledge, which validity is ascribed by distinct scientific communities, mode-2 science builds on problem-based multidisciplinary collaborations between universities, institutions, organisations and companies developing forms of knowledge primarily assessed by its contextual usefulness. The mode-2 theory suggests the emergence of a new mode of science where

> Society is able to "speak back" to science, and that this reverse communication is transforming science. Contextualization is invading the private world of science, penetrating to its epistemological roots as well as its everyday practices, because it influences the conditions under which "objectivity" arises and how its reliability is assessed. (Nowotny, Gibbons and Scott 2001: 54)

The way by which society 'speaks back' is conceptualised as a process of *contextualisation.* Although the mode-2 theory has effectively put on the science and innovation policy agenda the idea that such historical processes of contextualisation and transformation of science is taking place, the theory has been criticised for being rather vague in explaining what it actually means for the scientific process and dynamics of knowledge creation (Shinn 2002; Kamara 2009). Is 'contextualisation' just another word for stronger external steering of Science? Or might it imply epistemological changes in research cultures towards higher reflection on the societal context in which Science plays a role? Such questions still call for answers, and the field of public engagement is one way in which that might be further explored.

Historically the tradition for public engagement in Science and technology is often referred to as deriving from participatory technology assessment (pTA) as a further development of more traditional technology assessment, aiming to assess emerging technologies in order to provide advice on political regulation of these (Klüver et al. 2000). This very starting point implies that the tradition of public engagement has often been focusing on the political-regulatory level, rather than science-in-the-being. Accordingly public engagement is usually reactive rather than proactive; setting up ethical limitations to the application of science, rather than dynamically interacting with science itself. A slight disruption of this way of thinking, however, has begun. Correlated to the above discussions on the changing dynamics of contemporary science, the tradition for public engagement has started developing into the concept of *upstream public engagement* in science and technology (Wilsdon and Willis 2004). The concept has in particular been put forward by the UK think-thank *Demos* and has inspired European debates on public engagement. The key argument of this approach is that public engagement should not merely focus on the political regulation of the outputs of technological research downstream in the innovation chain, but also take place 'upstream' in the scientific innovation process:

> The task is to make visible the invisible, to expose to public scrutiny the assumptions, values and visions that drive science (…). [T]o enable debate to take place "upstream" in the scientific and technological development process, and not "downstream" where technologies are waiting to be exploited but may

> be held back by public scepticism brought about through poor engagement and dialogue on issues of concern.[2] (Wilsdon and Willis 2004: 19)

Obviously the ideas behind upstream public engagement have a lot in common with the thoughts of mode-2 science. Unfortunately, not least, by the fact that both are rather vague in describing how such dynamics of science might take place. So far only a few concrete experiments on upstream public engagement have been established, many of them reusing more traditional approaches, for example from participatory technology assessment in new contexts. The problem of doing so is that these methods are developed for a regulatory context, and not for upstream knowledge creation, thus implying only limited interference with the epistemological process of science.

If the ideas of mode-2 science and upstream public engagement are genuinely read as a new way to create knowledge, it seems that we are just in the beginning of exploring what the dynamics of such processes might be. Hence, it is the ambition of this book to further these thoughts, experimenting with new forms of upstream public engagement. The purpose of the book is to confront the societal role of science with the challenge of sustainability constituted by social environmental problems, and to make experiments on democratic deliberation in the field of upstream public engagement, in order to explore emerging potentials for new orientations of science, which more adequately reflects on and responds to these challenges.

2 Quotation from: HM Treasury/DTI/DfES, *Science and Innovation Investment Framework 2004–2014* (London: HM Treasury, July 2004), p. 105.

Chapter 2
My Research Interest

The research interest behind this book is multi-layered. It is a multidimensional expression of an action researcher engaging with the practical, theoretical and democratic aspects of furthering sustainability in its broadest sense. The Danish artist Per Kirkeby has portrayed the process of painting as a lengthy process of adding still more layers to a picture (Kirkeby 1996; Morell 2004). Every new layer added interferes with the canvas, and thus adding new colours might cover, blur or accentuate deeper layers of the picture. While the viewer of the final picture is only able to see the final expression of the artistic process, no layer that is added ever disappears from the canvas, but is part of the process towards the definite expression. I guess writing is a bit the same, except there is no final expression, only the epistemological limits of time, encouraging the writer to publish the work.

Articulating one's research interest is a complex matter. While overemphasizing this task might result in egocentric interpretations showcasing how the researcher's life history and the historic context encourages the research, too little focus on this matter leaves the reader with no transparent opportunity to judge the validity of the work. The aim of the following is to provide a balanced introduction to my motivations of writing this book in order to enable you to more fully understand what I try to say. It would be erroneous to argue that my research interest is not deeply grounded in my cultural context. Rather, it is based on some of the challenges I sense being part of contemporary society. A key motivation for doing critical research has been to avoid the tempting comfort of easily provided 'truths', which might effectively cover or blur the inconvenient condition that any insight is merely one tiny part of broader and more complex pictures. My intention has been to use this ambiguous starting point to seek to address questions which are either rarely asked or often simply taken for granted. In doing so I am strongly aware, however, that I am only able to do so by the imagination I have been provided through my cultural repertoire.

It is not my ambition to handle the issues of this book as abstract philosophical questions. Rather than thinking in stratified layers dichotomising the thinking-doing divide, I find that thinking and acting are interrelated and historically conditioned. It is through the altered combination of thinking and acting that the future will look different than the present. Accordingly, searching for alternative futures needs to combine conceptual imagination with the experience of doing.

It has been important to situate my work in one of the locus where these dimensions becomes visible: where science meets everyday life. In doing so, I have particularly focused on citizen participation and public engagement in science. I have practically and conceptually sought to challenge mainstream approaches in this field by confronting paradigms of public engagement with alternative forms of explanation found within the frameworks of action research. This approach to academic work methodologically links back to the process of painting. It is not merely through variations of deductive, inductive or abductive approaches that I am searching for new insights. As in the painting it is not least by the process of opposing, merging, stretching and sensing dimensions of paradoxes and consistency that this book has come into being. Doing so is not just a matter of creative, methodological experimentation. Rather it is based on the fact that we do not yet know what sustainable futures might look like, and rather than merely extrapolating our knowledge of the past into the future, this situation calls for imagination and experimentation. Not just for the case of methodological renewal, but because the epistemological approach needs to reflect the issue at stake. The aim of this book has not been to paint the perfect naturalistic picture within a given paradigm. Rather the motivation has been to search for cracks and openings in present rationalities by challenging these in any way possible. Consequently, I hope the final painting might inspire you in such direction.

Challenges in Upstream Public Engagement

My research interest in upstream public engagement particularly builds on the potentials of this area to connect scientific development with broader democratic deliberations on the role of science in society. Perceiving sustainable development both as a scientific and a democratic challenge, public engagement activities are a locus of particular interest. My interest has grown out of a continuous involvement in public participation in relation to science and technology, aiming to further new channels between institutions making decisions on socio-technological development and the wider public. My starting point, in the roots of critical theory, is that lack of public trust in science should not be explained simply by a public knowledge deficit, rather by the fact that the socio-technological development confronts and interferes with the lifeworld of ordinary citizens, creating new possibilities, new dilemmas and new problems. Hence, an important aim of public engagement is to enable ways in which the socio-technological development can obtain a higher correspondence with the human lifeworld. This is crucial because public engagement thereby holds a potential for connecting social and environmental aspects of sustainability. However, based on my practical involvement in this area, as well of academic examination hereof, a number of contemporary challenges appear clear to me.

Firstly, I find that the expansion of public engagement has increasingly been accompanied by strong interests in using these methods as social technologies legitimating predefined aims rather than inviting for wider public deliberation. 'Phoney participation' is increasingly invading the domain of public engagement, turning deliberative approaches into advanced governance tools.

Secondly, this dynamic enforces the ambiguity that public engagement per se reframes the opportunity for democratic dialogue in ways which are by no way straightforward: what on the face of it looks like furthering public dialogue might, by the framing of these dialogues and by rather complex social dynamics, obstruct marginalised public voices. Thus, public engagement is by no way innocent, it is a tool of power, calling for high level of reflexivity of those organising such processes.

Thirdly, despite the increasing number of innovative projects on science-society relations, it is far from clear that researchers involved in public engagement acknowledge public dialogue and lay peoples' experience as epistemologically relevant to their research. Rather, it seems, researchers often see public engagement as a means for communicating science to the public.

Fourthly, the attempt to invert this logic, by the approach of upstream public engagement, has so far mainly been dominated by rather traditional tools for participative technology assessment, leaving little focus on actually enabling two-way dialogue and inspiration between researchers and lay people. Accordingly, documentation of an evident impact of such dialogues on the upstream epistemological research processes has not yet been convincing.

Fifthly, working with public engagement in the context of sustainability has highlighted the need for working at societal level, by integrating social and environmental dynamics, and thus calling for the interference of local democratic and international scientific approaches. My community-based research in particular has highlighted a gap, between intellectual conceptualisations of sustainability and actual social realities of urban communities. Hence, taking into account real-world social dynamics seems as a focal point for actually making change happen, which is equally democratically and environmentally sustainable.

While technocratic rationalities might deliver the environmental adaptation needed to cope with climate change, they will hardly enable democratic restructuring of society at local, national or global levels. Without democratic approaches to the challenge of sustainability, it is not only the physical changes on Earth, but in fact also the potential future societal tensions grounded in social and environmental injustice, which might become decisive for our chances to sustain human life on Earth.

However, democratic answers to the challenges of social environmental problems are not just 'out there'. There is no way to simply deduct from the present state of unsustainability what future sustainability might look like. Rather, developing modern sustainable ways of living is a historical societal challenge calling for future development. The epistemological implication is that it does not make sense through public engagement just to ask for responsive public opinions. Doing so implies asking for a response to 'something', which in the case of future sustainability does not yet exist. Similarly, merely asking for public responses to technological development is not sufficient, as it implies that these innovations should themselves guarantee sustainability, which is contradictory to the conceptualisation of social environmental problems above.

Contemporary social environmental problems substantially addresses a *cultural* question, calling for scientifically *and* democratically to explore and develop sustainable ways of living. In my perspective we are just in the beginning of grasping how that might take place. Hence, in order, academically, to examine this question it is necessary to build on research methodologies suitable for these perspectives. For this particular purpose I find that the approach of action research offers valuable insights.

Action Research and Sustainability

Essentially what action research offers is a potential for developing a new field of practise responding to the challenges outlined in the above. Because the challenge of sustainability not least implies a cultural question of how we live in modern societies, calling for democratic examination and development, the question of participation moves centre stage. The particular relevance of action research is that enabling participation and social change is the very heart of this approach. To understand this in particular it is useful at this early stage briefly to introduce how I perceive sustainability from an action research perspective – and action research in a sustainability perspective.

Research as action, action as research. A first point concerns what it means to do research as *action* rather than merely observing, contemplating, analysing, abstracting and so on. The first does not oppose the later, but the element of action is another layer which transforms more traditional modes of research, and therefore the knowledge created in such processes. The argument for doing so is that to enable processes of sustainable democratic change it is necessary to understand and gain scientific insight by actually taking part in and experimenting with such processes.

Radical change transcending present realities. Concepts of 'change' are manifold: from minor to large scale, from bottom-up to top-down, from regulatory to emerging, from incremental to radical. By *sustainable* change I understand something which is equally radical and democratic. Acknowledging the unsustainable nature of contemporary modern ways of living, any sufficient change will be radical and transcend our present reality. For such change to be in any way democratic, democratic experiments transcending the present reality are needed. We need plural perspectives and people sharing various modes of knowledge and experience to develop sound alternatives to our current way of living. Technological fixes merely will not do; what we need is to identify new ways of living.

Enabling learning for social change. Sustainable change implies complex interrelations between environmental, technological, structural, cultural and social change. In addition to understand environmental aspects, social aspects of change need to be better understood and acted on. My epistemological claim is that the social must be part of the research process, in order for participants to gain insight by exchanging various forms of human experience. The span from the everyday life of local communities to overall structural decision-making needs to be unfolded, in order to encapsulate the dynamics making social and environmental change towards sustainability possible.

Nurturing trans-disciplinary understanding. For research in particular, this challenge implies ways of working which might appear somewhat different than traditional modes of science. To fully understand the interrelated aspects of sustainable change, it is necessary to transcend the research paradigms usually defining the scope for scientific thought (by trans-disciplinary I do not merely refer to inter-disciplinary activities, rather to processes *transcending* the paradigmatic defined limits of scientific disciplines). It is not merely through paradigmatic science wars, but also through trans-disciplinary approaches that multidimensional understandings of sustainable change can be interlinked. Working as a researcher, that is, being bound by the limitation of having to make particular methodological choices, which ambiguously enable/disable various kinds of insight, the challenge of sustainability calls for reflexivity, taking into consideration other modes of knowledge which are part of the bigger picture. A methodology focusing on change in particular needs to consider how to make such change work in reality for those it concerns. As the challenge of sustainability transgresses the limits of local/global, social/environmental, everyday life/planning, scientific/political, the potential blind spots and thus the demands for reflexivity are accordingly high.

When I highlight these features of action research it is not least to make clear an important ontological claim framing my research interest: that human understandings are always created and embedded in particular social practices. Hence to include human understandings, which are not only framed by research (or public engagement), but which reflect actual everyday life challenges of sustainability, these must be studied as part of their societal reality. Thus an essential methodological motivation behind this book has been, through action research, to move upstream public engagement from the academic arena onto societal arenas. Although acknowledging the global dimensions of social environmental problems, it has been the starting point for this book that solutions to such problems are always embedded in particular times and places, in particular practices and communities of actors. Hence, to include not only scientists and the arenas of research communities, but not at least citizens and their understandings of social environmental problems, I find that grounding democratic deliberations on future sustainability in the social realities of local communities is essential.

Research Questions for This Book

Based on the above considerations, the overall research question of this book is *how community-based action research in upstream public engagement can further new research orientations towards sustainability*. To analyse this question two intermediate sub-questions are addressed. First: *How are orientations towards sustainability created in a practice of upstream public engagement?* Secondly: *How can action research make a methodological contribution to upstream public engagement?* By working with these questions my intention is to add another layer to the discussion of upstream public engagement and possibilities for moving towards future sustainability, by exploring what is offered by community-based action research. My approach to doing so is to analyse the two sub-questions from a number of different perspectives. To do so, these will be developed into more concrete and theoretically, empirically and methodically grounded research questions. This is the aim of developing the research methodology of this book in Chapter 4. Before continuing this methodological discussion, however, the following chapter provides an introduction to the particular empirical starting point of the book.

Chapter 3
Case: Citizen Science for Sustainability

The writing of this book takes its empirical starting point in a specific initiative situated in the context of upstream public engagement: the Citizen Science for Sustainability (SuScit) project (Eames and Egmose 2011; Eames et al. 2009a; 2009b). The SuScit project was an initiative specifically seeking to provide local urban communities with a greater say in how priorities for environmental and sustainability research are defined, so as to ensure that future research more effectively addresses their needs. To do so the project methodologically aimed to develop new forms of collaboration between sustainability researchers, practitioners and local communities. Thus the project provides an example of upstream public engagement initiatives increasingly attached to setting research priorities and agendas for future research, and not least, for exploring the potentials and barriers in doing so. Methodologically, the project can be seen as hybrid between the approach of upstream public engagement and various inspirations gained from action research.

This book is written as a research project built on top of the SuScit project to undertake further analysis of the research experiences, and to explore in what sense these experiences of upstream public engagement can further the idea of new orientations of science relevant for meeting social environmental problems. My background for doing so is that I have been employed as a full-time action researcher in the research team developing, implementing and evaluating the SuScit programme.

Background, Purpose and Scope

The Citizen Science for Sustainability (SuScit) project was formed as a three-year London-based project aiming to develop new approaches to take into account community perspectives informing future research needs and priorities. In order to do so, the project methodologically aimed to develop a 'community-led agenda for urban sustainability research' as an example of how public engagement could help inform needs for future research. The project was funded by the UK Engineering and Physical Sciences Research Council under the council's Sustainable Urban Environment programme. In this context it was a novel initiative exploring new ways to inform the research council on societal research needs.

The project was situated in the contemporary field of UK research policy in a number of different ways. First, it was part of a long-term shift in the research council's strategic research from primarily focusing on wealth and economic development towards paying increasing attention to social wellbeing. Secondly, it reflected an increasing recognition of the need to find ways to further inter-disciplinary research. And third, the project was seen as a champion on what had emerged as a pressing research issue in the field of sustainability research: the challenges of environmental inequality and environmental justice. Research on these subjects had evidently showed a strong correlation between social and environmental inequalities as a real and substantive problem within the UK, highlighting that it is often disadvantaged members of society living in poorer neighbourhoods, who are being exposed to the greatest environmental risks, have the worst access to environmental goods and services and who experience the poorest health and quality of life (Lucas et al. 2004). Additionally these communities seemed to be the least likely to be engaged in dialogues about how science and technology can help to address these problems. Hence, to enable broad societal transitions towards sustainability, developing ways to better engage with and listening to the concerns of deprived urban communities, represented a substantial challenge. The project was formed in order to explore new ways to do so, and was in this sense a genuine experiment, a three-year methodological pilot study, to provide local communities with a voice in the future of urban sustainability research.

Project Set-up and Methodology

Within this setting, the project was set up drawing on the contemporary discussions of upstream public engagement in science. The outset for the project was a perception of sustainability as an inherently contested concept calling for open and reflexive framings. Rather than taking the outset in academic conceptualisations of sustainability, the aim of the project was to design a bottom-up public engagement process empowering lay citizens in dialogue with senior researchers and sustainability practitioners to articulate the environmental and sustainability research needs of marginalised and excluded urban communities. The aim of doing so was to find ways by which these participants could mutually learn from each other, thereby inspiring new research ideas and initiatives by including marginalised modes of knowledge and lived experience of urban communities.

This orientation was reflected in the overall project design, made up of two parts each containing a number of distinct activities. In the first half of the project local residents were engaged in community projects to explore and share experiences on what it was like to live in their local urban area. In the second half, the residents were sharing these perspectives with senior researchers and sustainability practitioners in order for them to respond by developing 'a community-led agenda for urban sustainability research'.

A Project Team, together with a number of external facilitators, and in dialogue with a Project Advisory Group, led the entire project: undertaking relevant background desktop research, methodologically developing and implementing the fieldwork program, and evaluating and disseminating the project. In order to address challenges of deprived urban communities, a fieldwork area was identified which met a number of indicative criteria: it was within the 5 per cent most deprived areas indicated by the UK Index of Multiple Deprivation; it was physically run down with local environmental concerns present and felt by residents; it had a large proportion of economically inactive local residents; and it had a centre suitable for recruitment of participants, prepared to actively engage with the research and having space to house some of the project activities. Thus, based on statistical data and a number of explorative fieldwork studies the project was set up in the Mildmay ward of Islington in North London, an area dominated by estates of mixed quality social housing with a high and diverse representation of economically inactive residents. Islington is a vibrant and dense urban area facing a number of complex challenges to urban sustainability. Even though it is the sixth most deprived borough in the UK (Islington Council 2006: 14–15), it is also home to some of the wealthiest people in the capital (Islington Strategic Partnership 2006: 9), and is an area of striking contrasts between affluence and poverty, facing significant social, economic and environmental challenges.

Figure 3.1 Social housing and Mayville Community Centre

Source: Eames et al. 2009. Reprinted from the Citizen Science for Sustainability project.

In order to establish the local groundwork and opportunities to continue the collaboration, the project was set up in close association with the local community centre. Mayville Community Centre, run by Mildmay Community Partnership, the Neighbourhood Management Partnership[1] for the ward, was the principal local venue hosting community activities including luncheon clubs for older people, youth clubs, IT, music, health and fitness classes, mother and baby and toddler groups, a community gardening club and outreach activities for the homeless.

The project was initially set up in close collaboration with the community centre by approaching local community group organisers. By presenting the initial ideas of the project these people became gatekeepers to the local community enabling the project team to explore what people found interesting about the project, who might be interested in taking part and, thereby, how the community part of the project could be framed in the local context.[2]

The actual recruitment of participants was based on inviting people to take part in a community film-making project about 'what it is like to live in the local area', whilst the issues of science and sustainability were gradually introduced later in the project, in order to give priority to people interested in addressing local issues. The implication was that most participants, rather than being attached to the science and sustainability agenda, joined the project either to take part in the film-making, the social activities, to get the bit of incentive payments offered or because they were already engaged in and feeling a responsibility about local concerns.

Based on criteria for social deprivation and the actual possibilities in the local context, three Residents' Panels were set up. The panels comprised mainly economically inactive residents, reflecting the ethnic and cultural diversity of the local area, and representing three distinct life stages in order to include different perspectives and lived experience. The 'Young Peoples' Panel' comprised 12 members (seven women and five men) between the ages of 16–21 years. The 'Women and Lone Parents' Panel' comprised 11 women aged 25–40 years. The 'Older Peoples' Panel' comprised 11 members (five women and four men) all over the age of 65.

---

1 The UK Neighbourhood Management partnerships are responsible on the ground for improving services, addressing service gaps and focusing action on local priority areas in disadvantaged neighbourhoods making them more responsive to local needs (Local Government 2011).

2 Whilst it was clear that this approach increased the sense of local ownership to the project, it also meant that the project primarily engaged socially active people. Still it seemed that these people had a broad experience on local concerns although they did not represent the socially most isolated residents in the area.

In addition to the Residents' Panels a 'Researchers' Panel' and a 'Practitioners' Panel' were involved in the project. The Researchers' Panel comprised 12 senior academics with expertise across a broad spectrum of urban sustainability issues (urban planning and design, transport, energy, water, waste, engineering, geography and so on). Covering a broad range of fields with relevance for urban sustainability research, priority was given to researchers who had a certain level of experience in framing research programmes, and were open to the idea and relevance of furthering community perspectives in sustainability research. Many of the academics had former research experience with community engagement and saw the project as an opportunity to further this field.

Figure 3.2 Members of the Older People's Panel

Source: Eames et. al. 2009. Reprinted from the Citizen Science for Sustainability project.

The Practitioners' Panel comprised 13 professionals from local, regional and national stakeholder organisations and institutions with expertise across a broad range of sustainability issues. The practitioners were invited to the project to include hands-on experience working with sustainability issues in practice and knowing possibilities and challenges in the field. Most practitioners were in senior and advisory positions having professional experience in community engagement. In this sense the members of the Practitioners' and Researchers' Panel were mostly professionals acknowledging the need to further community engagement and keen on taking part in a pilot project to do so. This choice of participants were made to further the possibilities to take the field of community engagement forward, keeping in mind the participants' prerequisites for doing so were probably above average.

First Half of the Process: Community Film-making

The aim of the first half of the project was to involve local residents in openly exploring and sharing what it was like to live in their local urban area. Rather than grounding the process in academic concepts, therefore, the purpose was for residents themselves to reflect on this issue. The key component in the community work, which was initiated by an initial focus-group meeting, was a number of community-led film-making projects for each of the three Residents' Panels separately. Over a period of two months, participants met on a weekly basis to discuss what it was like to live in the local area. The residents were skilled up by professional film-makers to do their own storyboards and shoot their own films reflecting the issues they wanted to address.[3]

The facilitation of the community projects was rather open. Rather than streamlining the process towards making films, priority was given to the participants to learn to know each other, feel comfortable with sharing their views and providing time enough to reflect on what they would like to address. Hence, it was important not just to reach for a specific output but also to create a social process for reflection and shared thought.[4]

3 A team of experienced university-based film-makers specialised in enabling marginalised groups to make their own films were facilitating the community film-making.

4 The dynamics of the three panels turned out to be somewhat different. For many of the young people the film-making activity itself and the incentive money offered was a key motivation. For the women and lone parents the social network of the group became of particular value. For the elderly reflecting on life and taking responsibility for the local area was particularly important.

Figure 3.3 Community film-making

Source: Eames et al. 2009. Reprinted from the Citizen Science for Sustainability project.

By the end of the community projects a total of 14 films were produced which in very different ways – dramas, documentaries, love stories – showed different community perspectives on 'what it is like to live in the local area'. But equally important, the three Residents' Panels had been working together in the groups, sharing their thoughts on local issues, which was an important for meeting the researchers and practitioners in the next part of the project.

Second Half of the Process: Shared Workshops

The aim of the second half the project was for Residents' Panels to share their experiences of living in the local area and for the Researchers' and Practitioner's Panels to engage in a shared dialogue in order to listen and reflect on how the community perspectives could be taken into account in furthering research and new initiatives towards sustainability. Through initial meetings with the researchers and practitioners these aims and roles were discussed and made clear. The deliberation process was primarily organised around four whole-day Shared Workshops involving all 70 people taking part in the project.

Figure 3.4 Shared Workshops. Visioning sustainable communities

Source: Eames et al. 2009. Reprinted from the Citizen Science for Sustainability project.

The First Shared Workshop was entitled 'Sharing Local Knowledge and Experience'. The aims of this workshop were to introduce all the participants, establish a positive and supportive forum and to share the residents' local knowledge, experience and different perspectives on the local area. This was done by showing the films and by facilitating shared reflections on the issues raised. Although the discussions were reported in written form, rather than aiming for a concrete outcome, the main purpose was to initiate a living process of sharing experience and perspectives across the three panels.

The Second Shared Workshop was entitled 'Visioning Sustainable Communities' and aimed to engage the participants in thinking beyond the present reality and imagining what the future could ideally look like in a 20-year perspective. The idea of reflecting on possible futures as the participants ideally would like them to be was an essential step to enable shared imagination transcending what is possible in the present. Based on the discussions in the First Shared Workshop a number of key themes covering the community issues raised had been identified by the project team. In this Second Shared Workshop the participants were invited to recall the issues raised by the community film projects and engage in a number of exercises specifically designed to foster a living dialogue about the issues across the groups.[5] Again the aim was to engage

5 Developing exercises to foster dialogue between community members and professional participants were a particular challenge in the project, implying difficult dilemmas in terms of the balance between making space for a free and open dialogue, and framing and syncing the multi-participant process; building on modes of communication equalling skilled professionals in dialogue with lay-people; and having time enough to establish a certain level of trust to share difficult issues, while also

in a fruitful dialogue sharing experiences on the issues rather than merely reaching for a concrete output.

Based on these two First Shared Workshops, the Researchers' and the Practitioners' Panel met at a two-day conference entitled 'Developing a Community Led Agenda for Urban Sustainability Research'. The aim of the workshop was twofold: for the participants to reflect on the perspectives brought up through the community engagement in order to develop ideas and proposals for specific projects, and to reflect on what would be appropriate inputs for a research agenda and a set of recommendations responding to the perspectives raised by the community.

Figure 3.5 The Researcher and Practitioner Workshop

Source: Eames et. al. 2009. Reprinted from the Citizen Science for Sustainability project.

At a final Third Shared Workshop these thoughts as well as ideas for concrete projects, which could respond to the local concerns, were reported back to the community. Through the project a wide range of ideas for further initiatives had emerged. Hence several dissemination meetings were held to support participants in taking these initiatives further on. Based on the inputs from these workshops the project team wrote up, in collaboration with the researchers and practitioners involved, a set of indicative research issues for each of the issues discussed in the project and a set of recommendations for future research on urban sustainability.

---

aiming for reflections and outcomes to be taken further. As in many public engagement processes this includes a danger of streamlining the facilitation process. Especially in the large Shared Workshops this problem became visible in terms of community members not always 'fitting in' to the process. While this from a facilitation perspective can be perceived as a difficulty, it is in an action research perspective an important indicator highlighting whether the process is engaging or overruling participants.

Reflections on the Outcomes

The SuScit project, in my view, fostered a whole range of outcomes at various different levels. As in any public engagement project some were very concrete whilst others by nature were more difficult to trace. Concerning the impact on research policy at institutional level the final project report, targeted specifically to the Engineering and Physical Sciences Research Council and similar funding bodies, included a number of recommendations in relation to sustainability research based on the experiences from the project. Although no formal institutional mechanism existed to take these recommendations on board, it was listened to with interest as inspiration for a number of new initiatives. However, it was also clear that this input was merely one of many agendas in research policy, and no direct actions were taken on the basis of the recommendations.

At practitioner level, the project inspired a number of different initiatives: at municipal level it inspired new approaches to sustainability and community engagement, at the local level it fitted well into the local community centre agenda of environmental refurbishing, and various other initiatives were established in continuation of the project. Probably much more practical action could have taken place if this had been given higher priority and funding.

Among the residents the process fostered learning and engagement, and several participants were interested in taking various activities further. For these people the project was seen as a genuine opportunity to make a difference concerning the issues brought up and fostering a lot of hope that something could be done. However, it also became clear that without firm, long-term support these initiatives tend to get bogged down.

For the researchers, the project turned out both as an inspiring experience, but also difficult in terms of working further on. This was directly reflected by the academics in the workshops addressing a number of fundamental barriers to further this type of community engagement. In particular it was found that:

> Building effective partnerships with local communities requires significant upfront investment of time and resources to establish contacts, build trust and relationships etc.
> – It is important to involve the local community at an early stage in developing shared goals for research.
> – The limited resources available to residents and local community organisations often limit their ability to participate in research initiatives from which they might benefit, or to which they might make a particular contribution.
> – The need to ensure that research also delivers practical benefits for community participants. Whilst research cannot necessarily deliver immediate solutions to sustainability problems, projects can seek to benefit community participants by building in education and skills development for those who take part.

> – There is often currently a mismatch in modes of funding available for research and what is required for facilitating effective community involvement, e.g.: i) the challenge of linking locally grounded small scale initiatives with large SUE consortia bids, ii) community participants or organisations may not be eligible to receive funding from research grants, etc.
> – Funding and researcher incentives: there is a need to provide greater recognition for non-academic research outputs.(e.g. outputs other than those published in peer-reviewed journals). (SuScit Final Report. Eames et al. 2009: 24)

Based on these various outcomes and reflections on the project, and from my own experiences and reflections being involved in the project, it seems evident to me that the key challenge to this kind of upstream public engagement was not so much to engage local communities in deliberations over local aspects of urban sustainability and the role science might play. The key challenge was twofold: to ensure that the deliberation process provided some tangible outputs for the community involved, and to overcome the barriers of academics and researchers to work in ways to enable such processes of change.

Whilst it is right to mention that the funding mandate and main purpose of the project was to develop a community-led *research agenda* for urban sustainability, rather than focusing primarily on delivering local outputs, and hence that different kinds of framing and facilitation could probably have strengthened the community dimension towards reaching more tangible outcomes, it is equally right that this challenge did not merely seem crucial for the project but also for academic work in more general terms. This particular issue was highly reflected through evaluation interviews with the participants: to work *with* communities to further *local* sustainability represents a genuine challenge to academic research cultures.

Building on these experiences identifying important questions to address in further research on upstream public engagement, the above challenges have been decisive for the writing of this book. In this sense the book is written as a response to the different groups of actors taking part in the Citizen Science for Sustainability project and similar kinds of initiatives: for the research community it is thought of as a methodological development articulating the potentials of action research in addressing challenges of sustainability. But the intention is also that these thoughts might appear somewhat rejuvenating to practitioners and people in power of framing processes of sustainable change. Last, but not least, these words are written in return to all those citizens sharing their often unheard insights, hoping that the thoughts of this book might inspire to take such perspectives into greater account.

Chapter 4
Methodology

Whilst this book builds on the experiences of a particular case, the Citizen Science for Sustainability (SuScit) project, it has been important also to establish a certain analytic distance to this project and develop a distinct theoretical framework setting the perspectives on the issues at stake. The particular value of such theoretical approach is that it holds a certain potential for making the invisible visible, to illuminate certain attributes of a more general order emerging from empirical experience. The danger in doing so is the easiness of simply extrapolating theoretical assumptions onto empirical experience. Doing so neither adds scientific value nor validity. Working theoretically with empirical insights addresses two principal challenges: first, the necessity to question one's own theoretical assumptions, as well as what empirically tends to be taken for granted, in order to address questions from which substantial *new* insights might emerge. Secondly, there is the high-wire act of substantially letting new insights emerge from the synergy between theoretical and empirical approaches. In the following I shall seek to make transparent my methodological approach to theoretical and empirical work, and in particular the interdependence between the two as the basis for gaining new insight.

Theoretical Grounding of Research Questions

In order to make the overall research question of this book operational, I am building on a number of theoretical approaches. My perception is not that any of these delivers the full explanation of the issues at stake, in fact no theory ever does. But I find that these theories, by reflecting the research interest of this book, make it possible to apply certain analytical perspectives and (referring to the analogy of the process of painting in Chapter 2) add new layers to understanding the issues at stake. Hence, the choice of theoretical frameworks is a process of deciding what layers and perspectives are relevant for examining the questions in mind.

In Chapter 2 the overall research question of this book was addressed: *How can community-based action research in upstream public engagement further new research orientations towards sustainability?* Under this overall heading two sub-questions were addressed. The aim of the following is to develop these sub-questions into operational research questions, which can be asked and further explored in the analyses and discussions of the following chapters. The first sub-question under this heading is the following: How are orientations towards sustainability created in a practice of upstream public engagement? Answering this question at least implies building on an explicit framework for understanding what is actually meant by 'sustainability'. To do so I have found it necessary theoretically to approach this concept rather critically, addressing two key aspects of this book: first, to take into account the paradox of modernity that scientific thought is not only offering solutions to, but evidently make up an inherent part of socio-technological dynamics, historically constituting present states of unsustainability. Secondly, the challenge of sustainability understood as building on social environmental problems: historical challenges of our time, which must be understood as both socially and environmentally grounded, and calling for scientific and democratic approaches. I find it essential to have a critical approach to sustainability, searching for an understanding which enables one to see more clearly the potentials and barriers at stake. To do so I employ a number of rather different theoretical approaches, each offering distinct perspectives on sustainability. These are: 1) Habermas' (1968; 1981) understanding of modern rationality, particularly attached to considerations over environment and sustainability in recent works by Elling (2008; 2010); 2) a critical perspective provided by Shiva (2005) contrasting European understandings of scientific and technological progress with a global outlook on sustainability issues; 3) an action research perspective on understanding social learning processes emerging from collaboration and lived experience (Nielsen and Nielsen 1999; 2006a; 2006b; 2007). Whilst these three perspectives all concern central aspects of moving towards sustainability, they are not part of a single conceptual framework; in fact they represent somewhat contrasting understandings, which I seek to combine into a critical understanding of sustainability. This is the particular purpose of Chapter 5.

Building on the development of this theoretical framework, and the case of Citizen Science for Sustainability (SuScit), it is possible to develop the first sub-question into an operational research question (1): *How can empirical orientations towards sustainability in the SuScit project be understood and challenged by theoretical concepts of sustainability?*

The second sub-question of this book was *How action research can make a methodological contribution to upstream public engagement?* This question reflects my own research experience, indicating that approaches of action research hold a potential for making a particular methodological contribution in this field. Analysing whether and how this might be the case, it has been necessary to build on a particular framework of action research theory. Taking into account the broad range of approaches to action research available today, I have particularly chosen to work with one specific approach, which I find adequately reflects and responds to the challenges motivating this book: the framework of Critical Utopian Action Research as conceptualised by Nielsen and Nielsen (1999; 2005; 2006a; 2006b; 2006c; 2007). Building on the observation that many public engagement initiatives only imply rather limited epistemological interaction between academics and lay-people, which in an upstream setting must be seen as a prerequisite for the dialogue to have impact on knowledge creation, it is essential to understand dynamics of learning and exchange of experience. While I am aware that the conditions for such processes are widely framed by institutional and organisation structures, I also find that to further the conditions for two-way dialogue it is essential to understand the epistemological dynamics of these. It is in this sense that I find the framework of *social learning*, grounded in Critical Utopian Action Research, of particular value. This had led to the research question (2): *How can the SuScit process be understood in terms of social learning?*

Empirical Grounding of the Research Process

In order to address the research questions, I work with a number of different approaches. Empirically, I am building directly on the experiences from the SuScit project. However, I have found it necessary to develop a methodological approach by which I build on three separate yet interconnected steps. The first step has been practically to design and run an action research process appropriate for exploring the dynamics of upstream public engagement in the field of sustainability. This was done though my work as an action researcher on the Citizen Science for Sustainability (SuScit) project. The second step has been empirically to engage with, study and understand the complex dynamics of this process in order to start exploring its challenges and opportunities. This was done as a separate research activity alongside the SuScit project. Although this approach obviously cannot be completely separated from the above first step, it adds another layer of reflection on the action research process. The third step has been, analytically, to examine and discuss the above fieldwork experiences in a theoretical perspective. These three steps constitute rather different ways of understanding. Still they are highly interconnected, and the core methodology of this book has been retrospectively to move across the experiences emerging from these, aiming to gain insight at a more reflexive level. Whilst some scholars often seem to prefer applying analytical methods constituted by the boundaries of a single solid research paradigm, the subject of this book in particular calls for the transcendence of silo-thinking, and for developing appropriate methodologies in order to more fully understand the issues at stake. Thus, in this book, these steps constitute distinct dimensions of understanding (or 'layers' if referring to the analogy of the painting).

The first dimension of understanding concerns the *facilitation* of the action research process. It is through the facilitation that the possibility to experiment concerning processes of social change emerges. This dimension is the practical essence of doing action research. Although an action research process is far from linear, the facilitation itself always includes framing and steering of the process. Accordingly, one of the blind spots of the facilitation perspective often is that the researcher does not pay appropriate attention to what is *not* happening and what alternative routes are enabled or not enabled through the process.

The second dimension of understanding adds an important aspect to the research process by an *empirical study* of the diversity of participant perspectives in the process. Whilst this cannot be separated from the facilitation perspective completely, the researcher can pay very different levels of attention and interest to this task (which is often highly dependent on the degree of instrumentalism and goal-orientation of the process). Being sensitive to alternative voices is thus important not only to the validity of the action research but also for the possibility to critically reflect on the fieldwork in a broader perspective.

The third dimension of understanding is to develop *reflexive theoretical interpretations*. The core idea of this third methodological step is to apply theoretical perspectives on social learning and sustainability as a framework of reflexive interpretation. Hence, this analytical dimension aims to highlight and understand *some* of the challenges and opportunities of upstream public engagement.

Building on the above dimensions of understanding, the core approach of this book is to apply three different yet interrelated perspectives corresponding to: A) first-hand experience facilitating the SuScit programme (Chapter 3); B) empirical fieldwork studies of multiple participant perspectives emerging from the SuScit process (Chapter 6); and C) reflexive theoretical interpretations of the SuScit project (Chapter 7). However, it is the methodological meta level of moving across the three different dimensions which enables the epistemological process of reflexively gaining insight from the SuScit process (an analogy could be to move across layers in a living multidimensional painting).

By moving from the first dimension (facilitation perspective) to the second dimension (empirical study), it is possible to identify tensions between the overall SuScit narrative and the diversity of participant perspectives on the process. This is crucial not only to understand the project on its own merits but rather on the multiplicity of participant perceptions on the process. By moving to the third dimension (reflexive theoretical interpretation), it is possible analytically to experiment with interpretations of these tensions in order to get a deeper understanding of their underlying premises by taking into account the other dimensions. Doing so implies building on particular theoretical framework(s); in this case, reflecting my research interest, exploring whether action research might be a way to further upstream public engagement.

It might possibly appear somewhat controversial building a scientific book on a methodology by which the analytical framework is used both for the interpretation of the fieldwork and for discussing the action research framework itself. How can you possibly compare two things if neither of them is fixed? The argument for doing so is that these dimensions are not stratified layers which can simply be added one by one; rather they are interconnected and co-existing dimensions of understanding the subject. Facilitation, empirical study and reflexive theoretical interpretation are analytic dimensions for understanding the subject, and the analysis of doing so is not to fix one or the other, but rather, by articulating these dimensions, to be able epistemologically to move across these perspectives as a hermeneutic tool for understanding. It is through the continuity/discontinuity of moving across these different dimensions that the insights and understanding of this book has emerged. Or, referring back to the analogy of the process of painting, the analytic methodology of this book is to enable the researcher to move across the painted layers on the canvas, exploring how colours in their various composition of co-existence interact with each other. Thinking across these dimensions of the research project is, in other words, not just a process of writing; it is an epistemological process of beginning to understand at a deeper level.

Epistemology Beyond Theory and Practice

A particular feature of my methodological approach is that it, rather than being framed primarily by *a* specific theoretical paradigm, methodological discipline or empirical field, has gradually emerged from working with a particular set of interrelated problems across these various levels. In the previous chapter I have already sought to outline this orientation. However, from a methodological perspective it is crucial to acknowledge this problem-based approach, building on the *principle of the object* by allowing the entire research framework to emerge from continuously and openly working with a particular set of problems. In this sense, the approach of this book is not merely inter-disciplinary (combining different research methods) but rather trans-disciplinary (transcending the boundaries of research methodologies). A principal for doing so is that the research paradigm itself should not determine the methodology. Instead the approach emerges from the field of study by continuously and curiously asking critical questions in order to understand the issues at stake and develop an appropriate methodology. In this respect the methodology of this book has grown out of a continuous multifaceted investigation of the research field over a number of years. It is this undetermined start for doing research which is the precondition for being sensitive to voices, dynamics and nuances in the research field, which might not yet have a corresponding academic explanation or appropriate paradigmatic methodology. If this approach might seem inappropriate to more traditional paradigmatic thinkers, that is the point exactly: as well as scientific paradigms allowing us to see certain things more clearly, it equally excludes other aspects. Trans-disciplinary modes of working are not a question of undermining the analytic strengths of science; rather they are a matter of walking the margins of science, listening to marginalised voices, critically exploring, identifying and re-addressing relevant research questions and developing appropriate research methodologies thereafter. Crucially the approach of trans-disciplinary methodology is far from opposing disciplinary modes of working. But it is, in fact, a question of insisting that the autonomy of the research is not merely constituted through the relation between the researcher and the research field, it is equally dependent on seeking intellectual freedom in terms of the opportunity to think freely beyond paradigmatically defined frameworks of understanding. A useful metaphor is that research is like looking through a prism which can be turned at different angles, enabling the researcher to look at the world from various perspectives. Applied at a methodological level this allows the researcher to triangulate different methods to obtain a fuller picture of the issues at stake. However, as all methods are deeply grounded in meta-theoretical assumptions, merely using methods ad hoc easily implies building on seemingly incompatible methods, which in an academic context is particularly vulnerable to the critique of pragmatism and methodological inconstancy. Accordingly, any multidimensional perspective must be firmly grounded in consistent meta-theoretical considerations in order to obtain scientific legacy.

One example of such methodological approach is the framework of reflexive methodology developed by Alveson and Skjöldberg (2000). According to reflexive methodology, various qualitative research traditions in social science should not (merely) be seen as competing paradigms; rather they enable the researcher to move across these disciplines to obtain a more reflexive level of understanding. Alveson and Skjöldberg show how empirical techniques, hermeneutical interpretative methods, critical theoretical analysis and post-structural approaches can be triangulated in ways in which the different perspectives of each method might supplement each other in delivering multi-perspective research insights. Taking my inspiration from Alveson and Skjöldberg, the particular aim of the methodology of this book has been to develop a methodological approach which through the research design adequately reflected the particular problems I am trying to understand. The methodology of this book is not a reflexive methodology in Alveson and Skjöldberg's particular use of the concept, but it is, as I have tried to outline above, a multidimensional methodology. At the end of this chapter I shall further outline the implications of this approach in terms of how I have chosen to outline the book. However, first it is worth elaborating a bit further on the methodological approach and in particular the epistemological presumptions and implications it has.

Possibly the above thoughts might appear somewhat intrusive to mainstream modes of academic thinking and classical understandings of the relation between the empirical and theoretical. Whilst scientific thinking often operates with concepts of induction, deduction and abduction as rational and relatively linear ways of explaining the relationship between the abstract and concrete, I have rarely met researchers stating that they fully follow the simplicity of such arguments in the research process itself. Rather the distinction between 'context of discovery' and 'context of justification' seems to cover an ambiguous relation between the actual epistemological dynamics driving researchers in their work, often rather different from the simplified linear rationality by which research-outcomes must be communicated in most journals and books in order to meet scientific incentive criteria (the same can be said about the writing of this book, which is consciously balancing between, on the one hand, obtaining academic legitimacy by following certain norms, whilst, on the other, not letting these determe the scope of thought). The often used stratified way of understanding the relation between the empirical (bottom) and the theoretical (top) is obviously a useful model in terms of constituting science as the only appropriate discipline producing abstract knowledge, but might be slightly misleading for actually understanding the relation between the two. Although stratified models on this dualistic divide can be traced through history all the way back to philosophic roots in Plato and Aristotle, alternative perceptions, however, have and do co-exist; take for instance Giordano Bruno developing a cosmology in which matter, spirit and infinity is one (and for which he was burned by the inquisition) (Bruno 1584), or Bateson's thinking on how spirit and nature are connected (Bateson 1979). Although the worldview of modern Science today seem incredibly dominant in its narrative, telling that the invention of Science was the basic precondition for developing modern society (a narrative that has been highly influential in social science as well), historical analysis of the actual emergence of Science shows a far more blurred picture of the way in which natural scholars have been influenced by extra-scientific historical conditions, perceptions and worldviews (Høyrup 2002). The purpose of briefly touching these historical perspectives it is not to go into comprehensive analysis of the philosophical roots of scientific thinking. Rather my intention is to highlight that intellectual thinking as it appears today is deeply framed by long historical and cultural traditions and influenced by present societal trends and interests; factors which more or less consciously and in ambiguous ways form the conditions for thought. Therefore I find it important continuously to address the question: What are the underlying frameworks constituting the way we think? Whilst it is contradictory to the above assumption to search for *one* such model (that would be ignoring the cultural diversity in this world), the epistemological consequence of thinking in these terms calls for experimentation and openness towards alternative ways of thinking. Accordingly, holding on to a stratified model of duality between theory and practice will have some or other effects on the way by which empirical data is perceived. In terms of the focus of this book to acknowledge lay-people's everyday life experience that could possibly be a devaluation of practical forms of knowledge being underprivileged to abstract academic thinking. In other words: to overcome the duality between lived experience and abstract thinking the researcher needs to be self-critical to her own more or less conscious ways of thinking, framing (or eventually determining) the way it is possible for the researcher to understand the subject at stake. While it is obviously not possible for the researcher to make completely explicit the cognitive structures framing her own thinking, it should be an obligation for any researcher at least to challenge the structures and frameworks constituting one's own epistemological process (which has been my motivation for writing Chapter 2 outlining my research interests).

Building on these considerations, an important inspiration for my work has been the idea of not dichotomising theory and practice by thinking in terms of 'what is general vs. what is specific'. Rather I find it more useful to conceptualise 'what is general in the specific'. This notion can be understood in a number of different ways. Flyvbjerg (1998) among others has shown how concrete case studies, without simply referring to linear logics of induction, can deliver qualitative insights inspiring theories at a more abstract level. Whilst this is an important step acknowledging the practical dimensions of science, it does not necessarily challenge the stratified relationship between the concrete and general. Another inspiration can be found in the academic tradition of hermeneutics making interpretations of a given subject through the circular movement between the concrete and the whole. Here the stratification is replaced by dialectic dynamic: it is through hermeneutical interpretation, moving between the general and the specific, that one can reach a higher level of understanding. A third conceptualisation of 'the general in the specific' can be found by actually transcending this duality, perceiving the general as embedded in the specific, and the specific embedded in the general. Such understanding can be found in the framework of Critical Utopian Action Research (Nielsen and Nielsen 2006), searching for 'what is general *in* the concrete'. Rather than the heuristic of induction, the metaphor of *listening* might be more useful for understanding this approach: it is by epistemologically listening to particularities that one might be able to *see* what has not yet been visible in the general (Nielsen 2010). In this framework of understanding, the particular is embedded *in* the general, as is the general *in* the particular (hence, you can take one tiny point of departure, start exploring how it is connected and continue exploring this world until forever). Thus the epistemological task for the researcher is not the inductive quest for abstract theories; rather to develop knowledge, which connects abstract thinking and lived experience.

When I am highlighting these three different frameworks of understanding the relation between the general and the particular, it is to make clear how pre-existing cognitive structures might impact on the way in which lay-people's everyday life experiences are perceived in academia, and that one of the challenges of this book therefore is at least to question such pre-understandings. The choice of Critical Utopian Action Research as an analytical dimension of understanding is thereby a selective choice in order to challenge – and hopefully enrich – scientific thinking and connect it to the practices of lived life. The outset for so doing has been to do my best to challenge my own understanding by trying to combine manifold ways of abstract thinking with several different types of empirical experience from the field. In doing so the imperative of this book has been: think freely! That is, challenging one's own assumptions, striving to really listen to the ambivalences of empirical experiences and let the ambiguities be the starting point rather than a barrier for knowledge creation. Doing so is far from easy. Making science out of it probably both requires systematic analytic abstractions and intuitively being part of the action research. This approach has a number of implications on how to analyse and explore the various sources of fieldwork experience. Films, recordings, written materials and so on are all 'materialised' outputs, but they are not merely products, but also *traces of social processes*. It is the clear assumption of this project that social processes are far more than linguistic: they are social, relational, processual, based on sense, experience and intuition. And so are the produced materials. If only analysing the project outputs through a purely linguistic approach, the understanding of the process becomes one-dimensional. Here a hermeneutic approach is needed, not aiming to explain but to understand things from the participant's perspective. This also highlights the importance of using various sources and inputs from all participants in the analysis. If the analysis is primarily based on the researcher's own perceptions there is a high risk of misinterpretation, not taking into account alternate perceptions and perspectives. As noted by Eikeland:

> We cannot just say; "I saw it", or, "I heard it", without *understanding* it; we are simply not able to articulate anything *if not* being able to understand and catch it. (Eikeland 2006: 203)

Indeed the nature the SuScit project, involving a high number of diverse participants in a joint collaborative process, was an incredibly complex process. Being involved in the project, simply trying to tell *one* story about such a process easily ends up with reproducing the intentions and organiser's narrative of the process, rather than actually focusing on the complex field of participant perceptions. The project was not just one but multiple different processes, with an even higher number of individual perceptions of what actually happened and what it meant to those taking part. Therefore, really understanding and telling the story/ies of the project is a highly complex matter. This has been the particular argument for developing the specific methodological and analytical approach into the particular structure of this book, which shall more concretely outlined in the following.

Book Structure and Research Methods

To reflect the above considerations this book methodologically builds on the combination of a number of different empirical, theoretical and analytical perspectives. Building on my research interest this book analyses the outlined research questions through the combination of the following chapters.

Chapter 3 provides an introduction to the Citizen Science for Sustainability (SuScit) project as seen from an organiser and facilitator perspective, reflecting the 'first dimension of understanding' (facilitation). The chapter in a relatively straightforward way aims to present the overall intentions and outcomes of the process, thereby providing the reader with a more transparent outset for understanding the later analyses. The facilitation of the SuScit process, being a hybrid between an upstream public engagement and an action research process, was designed on the basis of a number of best practices in contemporary public engagement, foresight and action research methodologies, rather than building directly on the framework of critical utopian action research. Still, some overlaps do exist, which provide the fundament for applying critical utopian action research theory on interpreting the potentials of the project. The chapter basically builds on the SuScit narrative developed by the project organisers through the process. However, to provide enough insight into the process to follow the later analysis, it has been supplemented by desktop research reports on research methodology, facilitation guides for the SuScit activities outlining intentions and methods as well as fieldwork notes and various different materials produced through the project.

Chapter 5 provides the theoretical framework for the analysis of this book. The chapter combines two different purposes: to discuss and establish a theoretical concept of sustainability (which is needed to address research question 1), and to provide a framework for social learning (needed for the analysis of research question 2). The theoretical examination thus provides a framework, which is continuously put into use throughout the book in order to better address critical questions concerning sustainability. The chapter on social learning aims specifically to provide an analytical framework for the analysis of social learning between project participants (Chapter 7).

Chapter 6 is an analytical chapter particularly aiming to reflect the 'second dimension of understanding' (empirical study), providing a hermeneutical analysis of the various different resident perceptions within the SuScit project. Reflecting that a particular aim of the project was to establish a community-based process, I find that paying particular attention to the community perspectives is crucial. Thus dedicating an entire chapter to focusing particularly on resident perceptions provides an important comparative outset for the later analysis of understanding the dynamics between residents, researchers and practitioners (Chapter 7). The chapter builds on a number of different types of empirical material: written outputs produced with or by the project participants; transcribed flip-overs, posters, notes, participant evaluations; transcriptions of taped focus groups and workshops; films produced by or in collaboration with the project participants; reflexive fieldwork notes and evaluations produced by project team members. The analysis presented in this chapter is built on combining these various inputs with a number of semi-structured qualitative evaluation interviews with the participants exploring perceptions of the project. Transcribing these interviews was important to get to know and seek to understand the perceptions at stake. Based on these interviews an explorative hermeneutic analysis was made to highlight participant perspectives on the project. This was an important step to make visible the many different perceptions at stake, and contrast these (not least with my own and organisers' perspectives on the project). From this basis three case stories were written to reflect central perspectives and ambiguities. An important part was, time and again, to go through interviews and materials produced throughout the project so as to contrast these different inputs and produce a text hermeneutically reflecting such key perspectives. The validity of this chapter thus builds on continuously contrasting the produced text with the various raw materials. This does not guarantee that every resident perspective is represented but it means that I have sought to take these into account. The text was further validated by being examined by the project team involved in facilitating the process with the residents.[1]

Chapter 7 is an analytical chapter building on the top of the hermeneutical insight of Chapter 6, adding the third dimension of understanding to the project (reflexive theoretical interpretation) by addressing the second research question: how can the SuScit process be understood in terms of social learning? A particular focus is to understand the relations between researchers

1 In terms of validity it would have been ideal if the residents themselves could have had a look through and respond to this material. Practically, this was not possible and hence continuously contrasting it with the various materials produced by residents (films, written materials, interviews) and examination by the project team was the second best choice.

and residents, not to ignore the value of practitioners taking part, but to reflect the particular research interest of this book. Thus the theoretical framework of social learning (Chapter 5) is put into use as an active interpretation of the SuScit project. This analytical approach is slightly different from the previous chapter in the sense that I am interpreting the project on the basis of a particular theoretical framework, rather than merely aiming hermeneutically to understand it in its own right. The analysis behind the chapter consists of two steps. The first step consists of supplying the resident interviews (used in Chapter 7) with a number of similar semi-structured qualitative evaluation interviews with the researchers and practitioners to make more transparent their different perceptions of the project. Again I have transcribed these to get to know and hermeneutically seek to understand the perceptions at stake. Based on these interviews an explorative hermeneutic analysis was made providing a comparative outset (including researcher, practitioner as well as resident perceptions). The second step was to make a reflexive theoretical interpretation of the dynamics of the SuScit project, in particular to explore aspects of social learning between the various participants. Doing so builds on addressing a number of analytical questions to the interviews, reflecting theoretical key aspects of social learning. Through the analysis it is explored in what sense this theoretical framework is applicable to the empirical experiences, both highlighting where aspects of social learning seems to take place, and where this was not the case. Hence, the purpose is not simply to state SuScit as a social learning process, but through this analytical perspective to understand certain features of the process. The validity hence builds on the hermeneutical process of interpreting the empirical experiences without extrapolating theoretical frameworks onto it. The produced analysis has been reviewed by researchers and practitioners involved in the SuScit project, and the final analysis of this process is what constitutes the chapter.

Chapters 8 and 9 provide a cross-cutting discussion connecting the various aspects and perspectives of this book. As already mentioned, the different dimensions of understandings are, although operationalised in separate chapters, highly interdependent and therefore call for an overall multidimensional examination of the research questions. In order to do so the chapter additionally builds on a critical discussion of the notion of *new modes of knowledge production* in order to put some of the challenges of upstream public engagement into a broader perspective. Based on these considerations, Chapter 9 provides the concluding remarks on the overall research question, how community-based action research in upstream public engagement can further new research orientations towards sustainability, and points out a number of perspectives for further action/research.

Chapter 5
Democratising Sustainability

The purpose of this chapter is to move towards a deeper theoretical understanding of the idea of sustainability. No theoretical approach can ever deliver the full concept of sustainability, or any other given concept. What a theory can do is to deliver a certain point of view from which we might be able to understand more clearly at least some aspects of the questions brought up for discussion. The purpose of this chapter is to conceptualise a theoretical framework, keeping in mind the overall research issue of this book, namely how should we understand how community-based action research in upstream public engagement can further new research orientations towards sustainability.

Let me make my starting point clear. The present global state of unsustainability, and the environmental crisis it implies, is closely connected to historical process of Western modernisation, now extending globally. Therefore, the challenges raised by unsustainability cannot be understood without examining processes of modernity. From this perspective sustainability it not merely a question of inventing greener technologies and developing more efficient economies, it is a basic human and cultural question of how we live on Earth.

My starting point for understanding sustainability differs from current mainstream Western approaches. Many approaches to sustainability, rather than addressing what seems to cause unsustainability in the first place, try to solve the challenge of sustainability by the very same means that have seemingly been part of creating the problem. One such example is so-called green capitalism, which argues that a new industrial revolution of increased economic and environmental efficiency will move societies towards sustainability (Hawken, Lovins and Lovins 1999). A second example is the framework of ecological modernisation which suggests that new modes of technological development, which take environmental parameters into greater account, will in themselves solve our problems (Moll et al. 2009). Whilst I fully agree that we need to rethink Western concepts of technology and the economy, I hardly think that it is possibly to do so, only from within the very same conceptual frameworks, which need to be transcended. I certainly agree with the approach of transition management that we need societal transitions towards sustainable modern societies (Eames and Egmose 2011). But I find it strongly problematic if we only ask the rather pragmatic question of what can be done within the framework of current systems, while ignoring the fundamental question of why unsustainability has emerged in the first place. As even world leaders apparently leave it to the future to decide whether incremental system changes will solve our global problems without addressing their fundamental cause and scope, I shall not put myself in a privileged position to finally judge whether this is in fact a substantial opportunity. However, rather than just waiting for the answer, I have decided in terms of my work with this book more fundamentally to seek to understand the problem of unsustainability. Consequently I am trying to rearticulate the critical edge of sustainability. The Brundtland commission defined sustainable development as 'development that meets the needs of the present without compromising the ability of future generations to meet their own needs' (World Commission on Environment and Development, 1987: 43). However, one could ask what is meant by 'future generations'? One generation? Two? Or infinite? The radical interpretation of sustainability would be not merely to take into account our children and grandchildren, but the future human civilisation as such. In this perspective, sustainable development is not a question of having resources for surviving the next centuries, but to enable modes of living that can ensure that humans can continue to live on Planet Earth. In this respect, the weakness of the concept of sustainable development (and possibly the reason behind its popularity in the Western world) is that it merges ideas of 'sustainability' and 'development' into one single concept. Sustainable development becomes a strategy for sustaining the Western society itself, taking into account and aiming to manage its impact on the environment (Sachs 2010). However, whilst development in this sense is most often interpreted in terms of economic growth, what is needed is to broaden our understanding of what is meant by development in a broader sense (Jackson 2009). To do so in this book I have preferred working with the concept of moving towards sustainability, putting the notion of sustainable development in a broader perspective.

Modern Western civilisation is one out of many historical cultures, which has populated Earth over time. What is it that makes this civilisation unsustainable? If addressing the challenge of sustainability in this overall perspective, one quickly realises that a greater cultural outlook is badly needed for understanding ourselves at this historical stage. If only addressing sustainability from within our current economic or technological horizons, there is a great risk of only seeing part of the challenge we are facing. To transcend the current state of unsustainability we need to transcend our own modes of thought. Taking into account the globalised scale of the problems, an obvious opportunity for such transcendence includes a global cultural outlook. We need to ask ourselves: what is it that causes unsustainability? And combine insights regarding this fundamental question with our knowledge about the way modern society functions. It is within such broader perspectives we might best be able to transcend the present state of unsustainability. This at least has been the motivation for writing this book.

Based on this motivation this chapter combines two rather different perspectives: the Indian thinker and environmentalist Vandana Shiva's understanding of sustainability, which is developed in her book *Earth Democracy* (2005); and the Danish sociologist Bo Elling's critical theoretical understanding of sustainability put forth in his book *Rationality and the Environment* (2008). The perspectives of these two thinkers are rather different. While Shiva's concept of sustainability concentrates on human relations to nature; Elling primarily conceptualises societal dynamics. While Shiva sees sustainability from a Third World perspective, Elling analyses institutions of modern Western societies. While Shiva shows how the modern Western worldview leads to increasingly unsustainable conditions in the globalised world, Elling analyses the capabilities of modern societies to further sustainability. Thus, combing Shiva and Elling implies operating across distinctions of nature/society, local/global, north/south.

My intention is not to make a synthesis of the two positions. To put it simply, the philosophical foundations of Elling and Shiva's work are too diverse to be merged into one consistent sociological framework. Rather my intention is, by contrasting these perspectives, to be able to more clearly address the challenge of moving beyond unsustainability. What Shiva offers is a critical Third World perspective on why modern Western societies are not sustainable. What Elling offers is a critical Western perspective on the dynamics of modernisation. By approaching unsustainability from these two different angles I hope to be able to make visible some of the paradoxes of sustainability.

In the following I shall present Shiva's conceptualisation of sustainability, followed by an introduction to Elling's analysis of the rationality of modernity. On the basis of these two perspectives I will build on concepts of *social learning and social imagination*, which I perceive as a relevant approach for overcoming some of the challenges addressed. In doing so I draw on the collaborative work of the Danish action researchers Kurt Aagaard Nielsen and Birger Steen Nielsen (1999; 2006a; 2006b; 2007). The chapter is rounded off, on the basis of these various insights, by conceptualising how I perceive the concept of sustainability as a cultural and democratic challenge.

Shiva: Unsustainability and the Process of Enclosure

Diagnoses aiming to understand relations between contemporary Western societies and concepts of sustainability are countless. Without ignoring this diversity I shall elaborate on one particular understanding, which I find useful for addressing the question the other way around: What is it that makes societies unsustainable? By looking at Western societies from the outside, we might be able to address what unsustainability actually is. This is what the Indian thinker and environmentalist Vandana Shiva has done.

Reminding us that the term economics derives from the Greek *oikos*, meaning home, Shiva argues that we need to perceive Planet Earth as one household, one multiple economy. In her book *Earth Democracy* (2005) she identifies three major economies which frame the dynamic of this household Earth. The *first* and without any comparison largest economy is nature's economy: the continuous material cycles and reproduction of the biosphere. Living organisms, animals and vegetables, water, soil and nutrients all take part in major interrelated cycles sustaining life on Earth. The *second* economy is the sustenance economy: the work of people providing the conditions necessary to maintain their lives. It is through the material and cultural processes of the sustenance economy that production and reproduction become possible. The *third* economy is the market economy. The market economy is not to be mistaken for the market as a concrete place. People of all times have had markets on which they have exchanged goods. What constitutes the market economy, driven by economic capital, is that it has become disembodied, dislocated and decontextualised. In the market economy, financial value has been decoupled from actual flows of goods. In the market economy, economic value can be created without correspondence with nature's economy and the sustenance economy.

To understand this Shiva uses the historical concept of enclosure. In her analysis of the emergence of modern Western societies she shows how the mechanism of enclosure has been transforming the perception of what were earlier seen as commons, the nature in which we live: Forests, water and field lands, which people were living in and with, on which they knew their existence depended and which were a common responsibility to sustain. Such areas were perceived as commons, constituted by local communities' ability of democratic governance as key to sustaining the co-existence of people in nature. Overuse of nature simply would undermine the basis of existence and lead to the extinction of people. This calls for social organisation. As Shiva puts is: 'A democratic form of governance is what made, and makes, a commons a commons' (Shiva 2005: 21).

A critical reading of Shiva implies raising the point that pre-modern governance of human-nature relations in several historical cases has been far from democratic. However, what Shiva essentially points out is that the human-nature relation changes with the emergence of modern Western culture, and that the concept of enclosure plays a key role in this transformation. Enclosure, in its simplest form, is the process of 'surrounding a piece of land with hedges, ditches or other barriers to the free passage of men and animals' (Shiva 2005: 19).[1] Through enclosure the commons are transformed into private lands. In

1 Quoting G. Elliot Smith in Rifkin, Jeremy: *Biosphere Politics*, 1991, p. 39.

her writing Shiva shows how this process historically took place in England in the seventeenth century onwards. Through enclosure nature was transformed from commons, providing the basis for co-existence of human and nature, to resources, something that could be used. The value of resources is estimated in terms of economic market value. But when the value of goods becomes decontextualised from its origin, people lose their lived sense of responsibility and co-existence with nature. Hence, the implication of the process of enclosure is a transformation of the human-nature relation. By enclosure the perception of value is transformed into assessments of economic market value. This mechanism is crucial to Shiva's critical understanding of the West:

> The transformation of commons into commodities has two implications. It deprives the politically weaker groups of their right to survival, which they had through the commons, and it robs the nature its right to self-renewal and sustainability, by eliminating the social constraints on resource use that are the basis of common property management. (Shiva 2005: 29)

Nature's economy and the sustenance economy, which were once seen as the foundation for human progress, are not even counted for as value in the market economy.[2] Therefore increasing efficiencies and great market economic growth in GNPs can happen on the basis of enormous losses of nature's economy and the sustenance economy. 'The key to the domination of the market economy is its ability to claim resources from outside its scope' (Shiva 2005: 19). However, Shiva reminds us:

> The illusion of efficiency is produced by externalizing the ecological costs (…). Economic growth takes place through the exploitation of natural resources. Deforestation creates growth. Mining of ground water creates growth. Overfishing creates growth. Further economic growth cannot help regenerate the very spheres, which must be destroyed for economic growth to occur. Nature shrinks as capital grows. Furthermore, while natural resources can be converted into cash, cash cannot be converted into nature's wealth (…). The increased availability of financial resources cannot regenerate the life lost in nature through ecological destruction. In nature's economy and the sustenance economy the currency is not money, it is life. (Shiva 2005: 32–3)

The key to sustainability in nature's economy and the sustenance economy is the continuous material and cultural reproduction of life. Shiva's answer to the

---

2 As Shiva notes, perceptions of the sustenance economy can hardly be separated from issues of gender and cultural depreciation of human activities associated with household work (often not even articulated as 'work').

question 'what is unsustainability' seems to be: the undermining of nature's economy and the sustenance economy as the self-regenerating foundation of life.

The implication of this way of thinking is that mainstream understandings of sustainability, in terms of equally balancing environmental, social and economic factors, appear rather insufficient (see Figure 5.1 below).

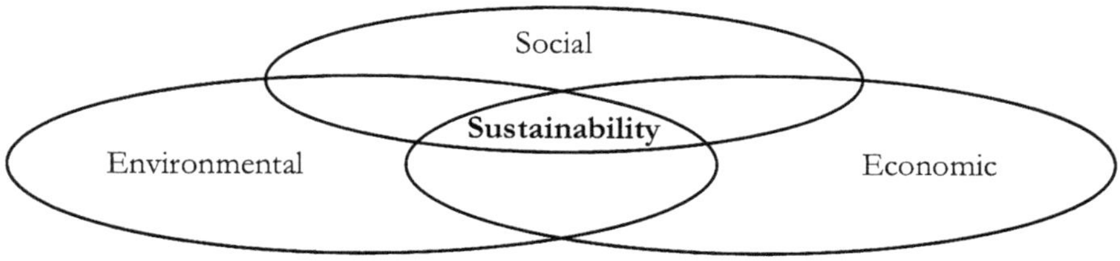

Figure 5.1 Sustainability conceptualised as a balance between environmental, social and economic factors

Rather Shiva prefers to visualise her conceptualisation in a different model, highlighting the unequal and stratified interrelations between nature's economy, the sustenance economy and the market economy, whereby she distinguishes between stable and unstable constellations of these (see Figure 5.2 below).

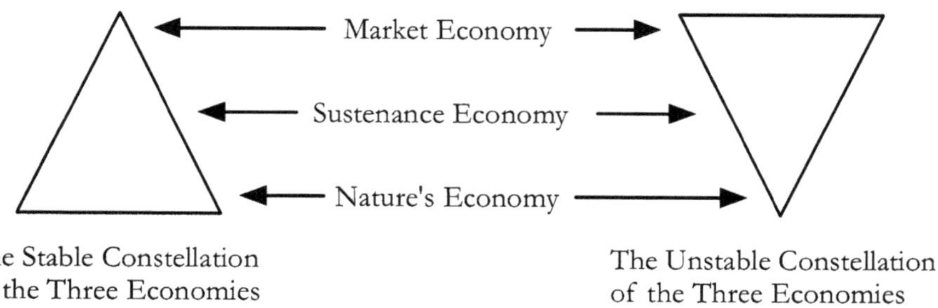

Figure 5.2 Shiva's model of constellations of the three economies

Source: Based on Shiva, V., 2005: 52, Figure 1: *Earth Democracy. Justice, Sustainability and Peace.* South End Press. Copyright © 2005 by Vandana Shiva.

In Shiva's conceptualisation, nature's economy is the foundation for the sustenance economy, which again is the basis for the market economy. Both the stable and the unstable constellation can include all three economies. But where the unstable state emerges from the exploitation of the foundation, the stable state builds on processes of self-renewal of nature's economy and the sustenance economy:

> Sustainable societies move in a stable state (...). To be in a stable state is not to be motionless (...). Contemporary ecology movements represent a renewed attempt to establish that steadiness and stability are not stagnation, and that balance with nature's essential ecological processes is not scientific and technological backwardness, but rather a sophistication towards which the world must strive if planet earth and her children are to survive. (Shiva 2005: 51–2)

Thus Shiva raises the fundamental question concerning the aim of household Earth, noticing that the key difference between nature's economy, the sustenance economy and market economy is that while the first two flourish through continuous material and cultural reproduction, the latter currently aims primarily for economic growth. From this point of view moving towards sustainability therefore not at least implies a cultural reconsideration of what we are really aiming for in the first place, and an implicit critique that financial growth, instead of being merely means, is falsely perceived as an aim in itself. Contrary to this perception, Shiva's understanding of the aim of civilisation would probably be: sustaining life itself without undermining its very foundation. Shiva points out that building on a stable sustenance economy has been the key to any civilisation which has been able to save itself from extinction. And she shows that the Western model does not lead local communities in the rest of this world in such a stable direction. Rather on the contrary, she argues through numerous examples that stable local economies have existed all over the world and could continue to do so if not hindered through the globalised dynamics of Western value systems.

While this critical perspective provides some good food for thought about the domination of imperialistic Western values in the rest of the world, it is less exact on how Western societies themselves can move in a sustainable direction. Modern Western societies are built on and depend on processes of social differentiation, the division of labour and expertise permeates all aspects of modern daily life. The historical processes of modernisation are not simply reversible. Therefore, taking into account Shiva's perception of what makes Western societies unsustainable, the question still persists, how to move in a more sustainable direction. Or in other words: How can we overcome that very dynamic which seems to ignore the fundamental value of nature's economy and the sustenance economy to the benefit of economic growth?

From Shiva's point of view, what has been lost in modern Western societies is the direct relation to, dependence on and therefore responsibility for nature, as well as people's own sustenance. How can such development be understood in terms of Western thinking? In fact, this is far from impossible; it has even been an essential object of study of many Western thinkers throughout the previous century (Nielsen 1997: 138ff). To mention a few: Habermas (1968; 1981) has critically discussed the emergence of expert culture as the instrumentalisation peoples' lifeworld, even of those aspects where instrumental rationality is clearly not suitable. Beck (1992) has conceptualised the emergence of the risk society as building on a paradoxical quest for certainty, which cannot be guaranteed by the increasing foundation on expert knowledge, because such knowledge is by nature never certain. Negt (1984) has, employing the term erosion crisis, exemplified how the social putty, which holds society together, seems to be eroding. Nielsen (1997: 144ff) has in his interpretation of Negt further argued how the erosion crisis can be seen as closely connected to the way expert culture colonises the human life world. In other words, the way in which Shiva states that nature and the social have been disconnected in modern life is far from extraneous to Western conceptualisations of the role of expertise; in fact the particular focus of this book on expert-layman relations is central to understanding processes of societal differentiation and modernisation. In the following, I shall try to address the question as follows: What are the possibilities within the field of expert-layman relations to transcend some of those challenges raised above? To do so, an elaborated understanding of the dynamics of modernisation is needed. This is the focus of the following section.

Elling: Instrumentalisation of Human Rationality

If the current state of unsustainability is closely related to the process of modernisation itself, understanding modernity becomes a crucial step to understand how challenges of unsustainability might be transcended. The key issue addressed in this chapter is: What has happened to human rationality in modern times? Understanding this question in the field of environmental and sustainability politics and governance has been the focal point of the Danish sociologist Bo Elling in his book *Rationality and the Environment* (2003; 2008). In this work Elling in particular builds on Habermas' *Theory of Communicative Action* (1981), analysing some of the paradoxes of modernisation. The focal point in Elling's work is to understand the nature of human rationality in a societal perspective. Whilst the word *rational* is usually associated with very particular meanings of what it means to act rationally, *rationality* in this particular sense is a broad concept fundamental to societal dynamics.

Building on Habermas' theoretical framework, Elling recalls that the historical process of modernisation has implied a change in rationality based on processes of societal differentiation:

> Modernisation is the process that transforms society from being reproduced by cultural traditions, norms and habits to a society characterised by reflexivity and a societal divide into autonomous parts, fractions or sectors. (Elling 2010: 31)

Societal differentiation has played a key role in the process of modernisation. Through differentiation, separate spheres of expertise, building on their own value systems and rationalities, have emerged. It would be erroneous to postulate that this process has not enabled an impressive increase in societal capacity, which has today become highly appreciated in the modern Western world: the ever growing number of fields of increasingly detailed expertise, the still more specialised roles and functions found in society, the expanding understanding of still more aspects of the world in which we live. The problem, however, arises because the substantial nature of this development has appeared to be rather double-edged, as originally captured in Adorno and Horckheimer's work *Dialectics of Enlightenment* (1972) and further developed in Habermas' work:

> Based on the process of differentiation, modernity can be characterised by two factors. First, the differentiation of value spheres of science, morality and arts to a professionalism that means an increasing differentiation between expert cultures and the broader public. Second, under the capital economy, modernization processes have been found to favour one above others of these expert cultures: the cognitive-instrumental. (Elling 2008: 101)

Concurrently with the increase in expertise-based capabilities the problem emerges that expertise does not merely add new societal capabilities; it also takes over fields of action and decision-making, which used to belong to other spheres. Thus modernisation is characterised by two parallel processes: societal *differentiation* and processes of *colonisation* whereby expertise increasingly impact on the human lifeworld and rationality. Actions, choices and decisions belonging in the everyday lives of people are no longer taken primarily on the basis of human life experience, but increasingly build on differentiated fields of expertise. Whilst this can on the one hand be seen as an impressive increase in opportunities to cope with all kind of aspects of life, expert knowledge on the other hand also implies a devaluation of lay-people's day-to-day life experience as a solid foundation for decision-making. An inherent part of modernisation is the impoverishment of the lifeworld, which is decreasingly being seen as valid for decision-making. The substantial reason that this historical process turns out as problematic is that the differentiated expert functions of the modern society are not necessarily fully rational. In fact, they primarily represent one specific aspect of human rationality, the cognitive-instrumental. Thus, when expertise colonises the human lifeworld, it implies the dominance of a very particular form of rationality, which is not necessarily suitable for all aspects of human life:

> The cultural impoverishment arises because an elitist differentiation of the expert cultures is taking place away from the communicative everyday practice. (…) In other words, one-dimensional societal modernization has come about due to structural transformation. This is partly caused by the fact that certain forms of action – that is, strategic actions that are based on cognitive-instrumental rationality, have dominated media steering and the differentiation of systems, the economy and the administration. (Elling 2008: 101)

To understand what is being displaced by expertise, Habermas' theoretical framework offers important insights. According to Habermas, a full concept of human rationality must be understood as much broader than merely cognitive-instrumental. To demonstrate this point Habermas turns to a central point of human life: an analysis of human communication (Habermas 1981). In his analysis he shows how our communication is based on discursive expressions, which validity is ascribed by referring to no less than three distinct realities: an objective reality, where statements must correspond with the factual surrounding world; a social reality, where statements must correspond with cultural norms for social actions; and a subjective reality, where statements must correspond with the individuals own thoughts, feelings and aspirations, to which only the individual has full access. For a verbal expression to be perceived as valid in human communication, and thereby to enable mutual understanding, it must simultaneously correspond with objective, social and subjective realities. It must both be true in an objective perspective, right in a social perspective and truthful in a subjective perspective (Elling 2008: 87). According to Habermas, truth, rightness and truthfulness are not merely features of verbal expressions but three distinct dimensions of communicative action, and thus aspects of human rationality. Habermas argues in *Theory of Communicative Action* that in the human lifeworld these three dimensions are integral parts of human rationality. What has happened through the process of modernisation, however, is that these dimensions of human rationality, for the first time in history, became clearly separated through the process of differentiation. Through the separation of technical, social and aesthetic expertise singular types of knowledge are established: cognitive-instrumental (natural science), moral-practical (social science) and aesthetic-practical knowledge (humanities). Whilst these three dimensions used to be an integral part of human experience, they are now separated into differentiated spheres of expertise. This is what has enabled modern society to transcend pre-modern boundaries of social norms, traditions and dogmas, led by the Enlightenment thinkers believing in the full potentials of human rationality.

The emergence of modernity is not least a question of believing in human rationality. The question is where that process has led us. Here Habermas points out that one problematic implication of the process of differentiation is that scientific technical-instrumental knowledge can be counted for as valid, as long as it is perceived as objectively true, even if it is neither socially right nor appearing as subjectively truthful. The differentiation of knowledge-spheres into separate validity-systems is double-edged. On the one hand, in a knowledge perspective, this historical development has enabled science to build expertise on separate criteria for validity suitable for the specific object of study; the objective, social and subjective reality respectively. In fact the increase in modern expertise can be seen as enabled by the definition of separate scientific paradigms ascribing certain norms, values and criteria for expertise. On the other hand, however, in a societal perspective, it must be noted that the power, influence and impact of the various differentiated knowledge-spheres on the modern society has been far from equal.[3] This is what has led to the criticism of expert-culture built on technical-instrumental rationality: knowledge which appears as objectively true, without necessarily being socially right or subjectively truthful, has become ever more influential in modern society. Whilst this does not necessarily appear problematic to the professional knowledge-cultures constituted by this particular development, it is highly problematic in a democratic societal perspective. In Habermas' terms, cognitive-instrumental rationality represented by the technical sciences has transformed into a societal ideology impacting on aspects of society where fully rational decisions cannot be taken only on the basis of technical knowledge, but must build on communicative action; the integration of objective, social and subjective world perspectives (Habermas 2005). This is what appears to be the Achilles heel of modernisation: that the ever-increased capabilities of knowledge cultures and expertise are inherently accompanied by a narrow perception of human rationality as the fundament for societal decision-making.

To fully understand the implication of the above perspective, Habermas' distinction between system and lifeworld must be taken into account. The human lifeworld is the world as we experience through our everyday lives. In the lifeworld, everyday human actions are coordinated through communicative action oriented towards mutual understanding (Elling 2008: 79). The system, in contrast, the economic and political-administrative system, is oriented towards fulfilment of preset goals by ever-increasing efficiency. To act rationally means to fulfil preset goals, and the way cognitive instrumental rationality is measured is by the efficiency of this fulfilment. In the system, steering is not

3 The so-called *science-wars* of scientific disciplines fighting on the differences between natural science, social science and humanities can be perceived as the consequent battle for power ascribed to the differentiated knowledge spheres.

based on communicative action towards mutual understanding but replaced by money and power as steering-media. The historical process of modernisation implies a rationalisation of both system and lifeworld. However the different steering-media applied implies very different forms of rationalisation: 'In short, rationally motivated mutual understanding within the lifeworld, maximization of profit within the economic system, and maximization of legitimacy within the administration' (Elling 2008: 197). To Habermas, system and lifeworld must be understood as interlinked parts of society. However, the process of modernisation has implied that the system is increasingly colonising the human lifeworld, and accordingly system-based rationality replaces communicative action. In Habermas' analysis, cognitive-instrumental rationality has become the dominant aspect of the way modern systems operate. Thereby the colonisation of the human lifeworld represents a historical process, whereby communicative action is being replaced by cognitive-instrumental rationality.

What is the implication of this theoretical framework in a sustainability perspective? In his analysis on environmental policy and governance, Elling is quite clear: the mainstream concept of sustainability has become systemic:

> The concept is systemic because it primarily aims to ensure the conditions for the continued existence of the modern economic system in which the negative impacts of the contradictions within this system (...) are abolished or compensated. (Elling 2010: 35)

If sustainability merely is a systemic challenge, the appropriate answer is to further the rationalisation of the system towards ever-increased efficiency. This is exactly what is articulated in economic terms by the paradigm of Natural Capitalism, and in a resource perspective by Ecological Modernization. However, the systemic approach to sustainability, in a critical theoretical perspective, implies a true paradox. What Habermas' analysis of modernisation shows is that the dominance of system-based rationality is twofold: it concurrently increases certain societal capabilities, whilst undermining communicative processes grounded in the human lifeworld, essential for human capacity to cope with the challenges we might face. Thus, a societal strategy of ever-increased system rationalisation, as the answer to the downside effects of modernisation, seems to leave society in a highly unsustainable process.

Perhaps a more balanced way of perceiving the challenge of sustainability is to take into account that society has historically developed into a present state, where both systemic and lifeworld-based processes are necessary to sustain the modern society. The consequence of modernisation is that society can no longer build on the lifeworld alone; the system has irreversibly become an inherent part of the way society is constituted. This perception of the current state of society leaves us in a slightly different perspective on the challenge of sustainability, which Elling captures very accurately: the concrete challenge of sustainability is a question of fundamentally rethinking the relation between system and lifeworld (Elling 2010). Indeed one could ask: What would be a sustainable balance between system rationality and the human lifeworld? Continuously addressing this question might lead towards unfolding what has been launched as *A New Agenda For Sustainability* (Nielsen, Elling, Figueroa and Jelsøe 2010):

> The criteria for a new agenda for sustainability must be that the systemic rationalities as well as those rationalities from the lifeworld are taken together and made the basis for actions which include systemic connections, primarily in the economy and the political system (...). We must open up the possibility for other rationalities than systemic rationalities, and we must open up the possibility for other orientations of actions than goal orientation. (Elling 2010: 39)

Addressing the particular research interest of this book, the above perspective represents a new approach to the role of expertise in modern society, which can be approached from two sides: the one to consider is how differentiated expert cultures might be able to partly transcend the rationalities on which they build. The development of interdisciplinary sciences might be seen as a systemic attempt in such a direction. The other is to explore novel ways in which communicative rationality grounded in the everyday life can obtain a renewed role as the foundation for societal development. Newer experiments on democratic innovations can be seen as lifeworld-based attempts to further such an approach. The one does not exclude the other. Rather on the contrary, one could argue that both are necessary ingredients for moving science in a, culturally speaking, more sustainable direction.

A useful step in order to make this discussion operational in the perspective of this book is to introduce the concept of researcher/expert reflexivity in terms of cultural assessments associated with choices in the function as researcher/expert. Here Elling offers an important distinction, namely reflexivity on the basis of the lifeworld (socially mediated reflexivity), and reflexivity with a systemic form of organisation (systemically mediated reflexivity):

> Reflexivity on the basis of the lifeworld may then be described as a process that aims to reach understanding of something – its origin and consequence – in the outer world, the social world or the inner world, in which this understanding is realized through a combination of communicative action, normative foundation and cultural legacies. The more reflexivity takes place against the background of a rationalized lifeworld, the more it will be based on communicative action rather than on normative and cultural legacies. (Elling 2008: 197)

The relevance of socially mediated reflexivity builds on the fact that not only the system but also the lifeworld has become increasingly rationalised through the process of modernisation. The rationalisation of the lifeworld means that our decisions must be based not on tradition, norms and dogmas, but on dialogue where different arguments can be tested through communicative action in order to obtain a higher degree of mutual understanding, based on what appears as objectively true, socially right and subjectively truthful. Reflexivity associated to the *system*, however, is rather different:

> [R]eflexivity with a systemic form of organization must be characterized as a borderline case that concerns only the outer world, and in which the teleological and strategic forms of action, with their corresponding forms of rationality are decisive – that is, displace the forms of action oriented to reaching understanding. In this case, communication is media-steered, and the reflexive processes are involved in the pursuit of the imperatives of the steering medium concerned – just as their underlying and one-sided form of action, they are arranged with a view to meeting a specific objective. (Elling 2008: 198)

The consequence in a Habermasian perspective is the following: systemically mediated reflexivity alone is not capable of pointing out future directions for societal development, only to aim for the fulfilment of goals preset by the system itself. In this perspective, human everyday life is of particular relevance, exactly because it holds a potential, on the basis of communicative rationality, to transcend the systemic. Whilst it would be clearly naive (and contradictory to both Habermas and Elling) to claim that clear distinctions between socially and systemically mediated reflexivity can be drawn, this perspective might help us understanding some of the paradoxes of modernisation, and not least: why the challenge of sustainability is also a democratic challenge. Beck states:

> The themes of the future that are now on everyone's lips have not originated from the foresightedness of the rulers or from the struggle in parliament – and certainly not from the cathedrals of power in business, science and the state. They have been put on the social agenda against the concentrated resistance of this institutionalised ignorance by entangled, moralising groups and splinter groups fighting each other over the proper way, split and plagued by doubt. (Beck 1994: 19)

After all, reconsidering the historical process of modernisation of Western societies in a system-lifeworld perspective on rationality appears not to be completely inappropriate. In the following section I shall discuss what I perceive as one of the more promising approaches for moving further on in such a direction.

Nielsen and Nielsen: Social Learning in Everyday Life

> One of the key features of modern society is that it must now learn how to respond to the often negative consequences of its own actions (Beck 1994) and also how to overcome epistemic constraints that currently-held assumptions impose on "seeing" and "doing" things differently. This implies that modern societies need to learn much more quickly, more effectively, and much more critically than societies in the past (…). This, in turn, requires citizenries willing to participate actively in democratic deliberations, and capable of learning collectively, with and from each other. (Bawden, Guijt and Woodhill 2007: 139)

Through the above theoretical perspectives, citizens' everyday lives appear as essential to sustainability. In the following, I shall elaborate on my understanding of these concepts in order to make explicit their particular relevance for furthering the sustainability discussion. The sustainability perspective often raises the question: why focus on everyday life? Is it not increasingly evident that modern everyday life is far from sustainable, and that what is needed is global leadership pushing the sustainability agenda forward? Is it not a bit naive to think that anything can be learned from a far-from-sustainable everyday life-style, which can inspire and further sustainability? Possibly, but on the other hand: How can we say that potentials of sustainability are not embedded in dynamics of everyday life? If sustainability is, as hinted by the above theoretical perspectives, a cultural issue, some might even ask: Is not the ability to sustain life itself a fundamental feature of life? Here I shall not judge whether the first or the latter is the case, but for now leave the question open for further exploration. However, it should be noted that it is not necessarily contradictory to argue that both strong decisions on structural changes and much higher societal capacity to learn and feed back from lived life is needed to further sustainability. What I am particularly trying to address in the following is how to strengthen the latter (which does not imply to ignore the first).

In theoretically grasping how this might happen, I have found it useful to build on the collaborative work of the Danish action researchers Kurt Aagaard Nielsen and Birger Steen Nielsen and the framework of Critical Utopian Action Research (1999; 2006a; 2006b; 2007; 2015). Nielsen and Nielsen provide a theoretically and empirically grounded conceptualisation of sustainability in close relation to everyday life and social learning in a societal perspective.

A first point for anyone searching for insights that can be gained from an everyday life perspective is probably to make clear that lifeworld perspectives are far from simple or unproblematic. Rather on the contrary, as pointed out by Becker-Smith (1982), modern everyday life is constituted as ambivalent, paradoxical and potentially full of conflicts, which individuals and collectives must seek to find ways to accommodate. Accordingly it is not simple to learn from everyday life. To explain why it is still worth taking the challenge, one needs to understand everyday life in a particular societal perspective. According to Nielsen and Nielsen (2006a; 2007; Nielsen 2010), everyday life is far from extraneous to understand the modern society. Rather on the contrary: the historical societal conditions of our time are embedded in our everyday lives; therefore the particularity of everyday life exposes aspects of the societal. Based on the cultural and societal framing of the human lifeworld, everyday life holds the potential for transcending individual perspectives and to start understanding societal dimensions of the questions at stake. The question, however, is how to obtain such insight, and this is where the term social learning becomes particularly useful.

Concepts of 'social learning' are used in manifold meanings and contexts (Wals 2007; Keen et al. 2007). Here I shall merely try to unfold the particular sense of social learning conceptualised by Nielsen and Nielsen. The crucial point in their conceptualisation of social learning, and which distinguishes it from other meanings of the term and thereby also demarcates its use, is that the very foundation for social learning is the human lifeworld as it appears in everyday life, framed in its historic societal context. Whilst many other theories approach social learning understood as related to systemic change within a given organisation or institution (Ramage and Shipp 2009; Blackmore 2010), Nielsen and Nielsen's concept of social learning is a societal concept grounded in the citizens' lifeworld. Thus, the epistemological horizon of social learning is not to make systemic changes (merely), but to enable shared reflection and change as a societal dimension, which might counteract systemic rationality. Nielsen and Nielsen are rather critical towards more pragmatic uses of the term social learning. If social learning is only perceived as mediation of discourses (Dryzek 1997) or discursive reorientation (Pålshaugen 1998), the concept might be highly useful for systemic crisis-management, but not very sufficient for transcending systemic rationality. Or, as they put it: within the power-dynamics of mainstream roundtable dialogues vague voices grounded in the lifeworld are far too easily overruled by the power of established discourses of the system (Nielsen and Nielsen 2007). Thus, if a major democratic challenge of sustainability is to be able to learn from a lifeworld perspective, different approaches are needed. This is the background for Nielsen and Nielsen's development of their concept of social learning.

In their theoretical grounding Nielsen and Nielsen build on Mills' (1959) concept of sociological imagination – the ability to make changes in perspectives and link across seemingly heterogeneous ways of understanding – and they follow the transformation of the dominant cognitive orientation of this concept into Negt's (1975) idea of exemplary learning: the potential of gaining societal insights from even the most particular experiences. With Negt, the learning perspective becomes grounded in peoples' everyday lives, and the sociological imagination is ascribed as a basic capability of all people to make coherence in the ambiguity of lived life. Taking into account Becker-Schmidt's (1982) insights on the inherent ambiguity of everyday life, this task is far from simple. From this outset, however, Nielsen and Nielsen address the question: How can social learning take place grounded in the epistemological horizon of peoples' lifeworlds, and further, how can such learning processes involve not only people in their everyday life context, but also researchers and other professionals, in ways which enable both parts to mutually learn from each other? Exploring this question Nielsen and Nielsen build on a strong basis of empirical work developing bottom-up participation processes, not least the Future Creating Workshop. The orientation of this work has been to explore how various kinds of expert knowledge can meet and conjoin with everyday knowledge, in ways where expert knowledge, on the one hand, is not automatically seen as practically privileged to everyday knowledge, while on the other hand everyday knowledge can epistemologically neither claim to be absolute or 'true' (Nielsen and Nielsen 2007: 31). Thus their work builds on a particular perception of everyday life:

> Human doings and actions are not merely functional, but always manifestations of life as well. They are always embedded in existential situations, moments of our life, which have their meaning, exactly because that is what they are: moments in life. (Nielsen 2010: 33, *my translatio*n)

Aspirations of life are, although often hidden or disguised, embedded in lifeworld-based expressions. Lifeworld perspectives imply both knowledge and hope, not merely knowledge. They are based on lived experience pointing back and forth in lived history, by implying, connecting and reflecting critiques of the past and hopes for the future. This is where knowledge and experience differ. 'A warranted hope presupposes knowledge, insight and acknowledgement, but in no way a "complete" knowledge and acknowledgement' (Nielsen and Nielsen 2007: 30). However, while experience and hope is a fundamental part of human life, it is often also highly sensitive for individuals to share with others. This, in fact, is a focal point of Critical Utopian Action Research: to enable *free spaces* where critiques and aspirations embedded in human lived experience can be expressed and shared more freely than often is the case. To conceptualise the potentials hereof, Nielsen and Nielsen introduce the concept of *social imagination*:

> Social imagination is not the same as sociological imagination. While the latter refers to a critical intellectual cognitive dimension, social imagination is related to the gesture of performed and unfolded *life aspirations*, concrete ideas about, how one could live (…). The formulation of potential life aspirations are

> characterised by *pre-empting a different – a better or fuller – life.* (Nielsen and Nielsen 2007: 31–2)

Social learning on the basis of everyday life is not merely a process of creating knowledge or insight; it is constituted through a social process rooting the orientation and approach to societal questions, in the ground of citizens' lived life. Thus, social learning might offer an approach to overcome societal challenges inherently grounded in dynamics of modern institutions, because social learning as a democratic process holds the potential to overcome challenges associated with the systemic rationality. As underlined by Nielsen and Nielsen:

> In our imagination we can open up the possibility for action and change as possible answers to difficulties. But in our everyday life, we do not very often have the possibility systematically to develop social imagination. Action Research can create arenas – we call them "free spaces" – in which social imagination emerges in an easier and more productive way than what is possible in the structure of everday life. (Nielsen and Nielsen 2006a: 79)

This is where the concept of social imagination implies a more radical take on social learning. Social imagination is not merely a question of knowledge exchange. Since social imagination implies aspirations for the future – start imagining alternative ways and concrete ideas for how one might like to live – it additionally calls for a sense of responsibility to think in ways to act upon the issues brought to the table. In this sense social learning is not merely about creating insight and reflexivity. The demarcation of social imagination is whether new imaginative horizons emerge, not just for understanding, but also for actively relating to, and developing a sense of obligation to act or respond upon the problems brought to the table.

However, such processes do not emerge automatically. There is no way to enforce social learning to take place. But there are, at least, a few premises for social learning to take place. According to Nielsen and Nielsen, social learning is basically a question of how we meet the world. This is where concepts of researcher roles (Westlander 2006, Brydon-Miller 2008) moves centre stage for understanding the potential routes social learning processes might take. To understand researcher roles, and essentially researcher subjectivity, Nielsen and Nielsen introduce the concept of *free distance* as a prerequisite for social learning to take place:

> Only by positioning ourselves *in relation to* the world that is creating *a difference*, the interaction becomes a meeting (…). Roughly speaking, meeting the world is hindered or neglected, both in its epistemological and existential dimensions,

> in two ways: Either is the relation to the world characterised by far too big a distance or by absence of distance. The relation to the world based on mastery (of nature) – by which the world gets characterised as a *surrounding* – is carried by an arrogant and anxious gesture in relation to what we confront, and thereby precludes itself from befriending and thus be able to *understand* it. But the mirror image to this, the supposedly immediate unity or conflation with the world, is not any freer. The meeting as *being-with-the-world* is, as a third opportunity, characterised by a different gesture, namely tentatively to put oneself in the place of others, in place of the other (…). That is what can be described as a *free distance*. The formula for this meeting is *hermeneutic*, "question-answer-question": What do I meet, what questions does it raise for me; what answers do I have the possibility to respond with; what new questions emerge, become possibly etc. (…) Free distance thus is a key category. To be *in* the situation and *in relation* to it at the same time. (Nielsen 2010: 30, *my translation*)

Free distance is a necessity for social learning to take place. Therefore social learning is not a question of dialogue merely, but about our way of meeting and being-with-others and the other: to be able to be *in* the situation and *in relation* to it. In this perspective, sociality is the sense of the societal *in* the particularity of everyday life (Nielsen and Nielsen 2007: 33). Through their conceptualisation, Nielsen and Nielsen make explicit that social learning does not just emerge through any 'social' process. Social learning is basically an epistemological process and not merely discursive. Therefore for social learning to occur, more than dialogues and words are needed. Social learning is a social process between human beings, rather than a discursive meeting between lay-people and experts.

Through this conceptualisation of social learning, Nielsen and Nielsen offer at least one approach for transcending what Elling has named as systematically mediated reflexivity on the basis of mutual learning processes grounded in people's everyday lifeworlds. It is clear that such processes do not happen automatically. Therefore working with social learning must also be an empirical project. *Is* it possible to enable mutual learning between citizens and experts, and if so, can such process provide anything useful towards furthering sustainability? This is an underlying question for the empirical analyses in chapters 6 and 7. Before addressing this question, let me briefly encapsulate in the following what I think can be learned from the above theoretical insights of Shiva, Elling, Nielsen and Nielsen.

Sustainability as a Cultural Issue

What can be learned from all the above insights? As already mentioned the philosophical foundation and scope of these theories are too different to simply merge into one coherent framework. However, this should not prevent us from learning from these perspectives in order to get a fuller understanding of the concept of sustainability.

Although (or maybe because) Shiva builds on a Third World perspective, she offers at least two important insights within the particular scope of this book. The first challenges the way many Western perceptions of sustainability today articulate the environment, the social and the economy as horizontally balanced and equalised parts of a sustainable system. In response Shiva claims that in reality this relation is not horizontal but vertical; nature's economy is the substantial foundation for the sustenance economy being the precondition for the market economy. This relation can only be seen as horizontal on the basis of a 'rhetorical trick of calculation' not factoring nature's economy and the sustenance economy into the formalised market account. Thus both positive and negative transactions between the three are made economically invisible, although they might in reality be fatal. Shiva's articulation of the vertical rather than horizontal interdependency between the environment, the social and the economy rearticulates and reminds us in a very exact way of the critical edge of sustainability.

This leads to Shiva's second important point that unsustainability is highly related to the conceptual articulation of the sustenance and market economy as independent and de-coupled from nature's economy. Shiva's claim is that sustainability depends exactly on the way human actions relate to the environment, the social and the economy. In Shiva's perspective sustainability emerges from peoples' direct relation to and interdependency with nature. Because we *are* nature, we must live *with* and *in* nature in ways which does not erode the foundation of our own existence. Accordingly, as long as we treat the environment, the social and the economy as independent segregated subsystems, we are trapped in our own conceptual frameworks. Sustainability is by nature a coherent relational concept. This is why sustainable development is a wicked problem in the context of modernity.

In continuation of this critical perspective, Elling offers important insights into the dynamics of modernity, which might help us to apply Shiva's arguments in a Western context. A fundamental feature of modernity is the increase in the societal capabilities through processes of differentiation. In Habermas' terminology the system is a substantial part of making modern life possible, but the system itself has also clear limitations, because the system itself can hardly transcend the systemic rationality on which it builds. Thus, in parallel to Shiva's argument that the human sustenance economy can only exist on the basis of nature's economy, Elling could further argue that the legitimacy of the system must build on the human lifeworld. This perspective, however, raises a fundamental question, which needs to be addressed as contemporary approaches to sustainability: to what extent can the system establish and guarantee sustainable development? My perspective is that if the system is primarily self-referential and enclosing itself in its own rationality and enforcing this dynamic on its surroundings, it cannot. Sustainability is relational. Sustainability is not something that emerges from the system. It is an emergent ability of ecological and social *life*; to renew, reproduce and sustain itself. However, this ability can either be hindered or supported by the system. In modernity, differentiated expert functions have become inherent and constitutional parts of the continuous historical re-creation of society. For this process of modernisation to be sustainable it must build on a rationality, which respects and supports the ability of ecological and human societies to adapt, reproduce and renew themselves. This is a substantial essence of life, which cannot be compromised if we want to move towards sustainability. This perspective, however, represents the major systemic challenge of transcending instrumental rationality. What can be learned from these theoretical insights is that a key challenge for modern societies moving towards sustainability is to enable processes whereby socially differentiated expert systems continuously transcend their own rationality.

This particular challenge has been my motivation for bringing in Nielsen and Nielsen insights on social learning in a societal perspective. The essence of this approach is that because our lifeworld is embedded in a particular historical and societal context, everyday life exposes, in ambiguous ways, societal aspects of modernity. Everyday life experience contains more than system rationality. Applying Elling's terminology to Nielsen and Nielsen's approach, socially mediated reflexivity holds the potential to transcend the rationality of the system. Thus, if the rationality of the system has become an inherent part of the challenge of sustainability itself, learning from everyday life might be crucial. In the context of this book, Nielsen and Nielsen's experiences show that a free distance is needed for experts and citizens to mutually learn from each other. To be able at the same time to be *in* the situation, and in relation to it. Mediation between, what Elling terms as systemically and socially mediated reflexivity respectively, cannot be a discursive process merely. It must be a social process: a process of social learning, by which the particularity of people's everyday life experience are shared and brought into a societal perspective.

On the basis of the above insights, furthering sustainability implies the following: *First*, moving towards sustainability is not a question of merely inventing new technologies. Sustainability in its most basic sense is an immanent and emergent ability of ecological and social life continuously to renew itself. No technology can guarantee that ability, but technology can, as any other human invention, either support or erode that basic ability. Sustainable innovations are such who support processes of continuous self-renewal based on processes of life. The key challenge of sustainability is to ensure that human activity does not erode nature as the fundament, which humans themselves are part of and their very existence depends on.

Second, the centre point for sustainability is the essential human activities of everyday life. Everyday life is not something extraneous to sustainable transitions, it is the exponent revealing whether modern structures supports citizens in living sustainably. Everyday life itself does not automatically lead to sustainable transitions. But neither do societal transitions lead towards sustainability, if they do not support the self-sustaining capabilities of the life of citizens. Because sustainability is an immanent societal capability, everyday life echoes whether we are moving in a sustainable direction. Sociological diagnosis of cultural erosion of society must be taken seriously, because they reveal where the capability of self-sustenance is decreasing under current societal conditions. In this way everyday life is an important indicator for understanding unsustainable aspects of modern societies.

Third, in modern, socially differentiated societies crucial aspects of the human-nature relation are mediated and framed by institutionalised expert systems. This challenges whether modern institutions are actually capable of managing the human-nature relation, or rather functions on the basis of self-referential system-mediated reflexivity. The challenge of managing the balance between nature's self-renewal and activities of human sustenance requires processes of continuous feedback and mutual learning, taking into account both environmental and social aspects. Thus, the implication of social differentiation is that modern institutions must help facilitate the balance of the human-nature relation through processes of mutual learning. However, processes of mutual learning are only possible if institutions are continuously open to transcend the system-mediated reflexivity on which they build.

Fourth, the implication of the above challenge of modern institutions is that sustainability needs to be seen as a democratic societal project. Expert institutions have a crucial role in mediating between environmental and social perspectives, but they can only do so if both nature and society are perceived as having the immanent and emergent capability of sustainability. Expert inventions alone cannot guarantee sustainability, but sustainable innovations might be crucial to supporting nature and society in its continuous interdependent self-renewal. Thus the role of experts changes from being drivers of innovation towards supporting the sustenance of nature and society. This is in fact an appropriate sustainable equivalent to recent years' debate on rethinking science: the transition of scientific thinking into modes of thought, the very aim of which is to sustain society and nature, rather than systems themselves.

Is a Reorientation of Sustainability Possible?

Across the various perspectives presented above, one underlying theme emerges as increasingly important: the need to strengthen the relation between humans and nature. In the modern differentiated society, direct relations between people and nature have widely been replaced by societal structures. In modern life, nature is decreasingly a focal point of everyday life because institutions have been given the responsibility of managing, and indeed often intending to master, the human-nature relation. This implies an institutional paradox: when the human-nature relation is no longer a crucial aspect of everyday life, peoples' sense of responsibility for nature tends to turn over to societal institutional responsibilities. However, if the rationality of modern institutions primarily builds on reflexivity on a systemic background, the basic democratic human question of how we live in this world is widely constrained to an instrumental question of how to meet preset goals. This leads to the paradox that modern democratic institutions cannot navigate without a democratic mandate, but at the same time the public democratic capability to take responsibility for, discuss and make decisions about the human-nature relationship erodes when institutional expert cultures are perceived as having the final answer to social environmental problems. This leaves us with a major institutional challenge: How to ensure that modern institutions and expert cultures do not widen the gap between nature and people's everyday life? Most approaches to sustainability, despite all their good environmental intentions, build on systemic rationality, developing systems of expertise and furthering differentiation. In Shiva's perspective, however, this does not provide much attention to the fundamental historical mechanism of enclosure. Social differentiation has without any doubt enabled many societal capabilities. But it has also created a greater distance between the immediate lifeworld perspectives of people's everyday life, and their direct relation to and responsibility for nature as well as the social world. Sustainability is not least a question about responsibility. How can responsibility for nature and the social be maintained in a culture, where increasing social differentiation, specialisation and technological approaches are the answers to almost any challenge of sustainability? What I shall do in the following chapters of this book is empirically address this question in terms of whether it is possible through meetings between expert and lay cultures to approach the above challenges differently. I shall do so by addressing two questions through my empirical analysis of the Citizen Science for Sustainability (SuScit) project: first, what perspectives emerged through the residents' everyday life perspectives? (This is a focal point for Chapter 6.) And secondly, how were these perspectives understood and taken into account in the collaborative process between citizens and experts? (This is a key question for Chapter 7.)

Chapter 6

Unfolding Everyday Life Perspectives

The aim of this chapter is to unfold citizens' everyday life perspectives which came to light in the Citizen Science for Sustainability (SuScit) project. The purpose is not to present a full list of subjects or to provide a complete account of the themes that were discussed. Rather, the ambition is critically and hermeneutically to understand what was at stake for the residents involved. My reason for paying particular attention to everyday life perspectives relates to the argument presented in the previous chapter, namely that to get a real understanding of what sustainability might be, it is insufficient merely to examine academic conceptualisations.

My approach is critically hermeneutical: on the one hand my aim is hermeneutically to understand what was at stake as seen from residents' standpoint. On the other hand I am also as a researcher playing an active part in critically interpreting these perceptions, questioning them and seeking underlying orientations. Thus I am building on the presumption that the inherent potentials of the community dialogue are not always the strongest discourses articulated loudly and clearly. Sometimes important ideas might appear rather ambiguous and vague at first sight. As an active interpreter it is my role to listen to vague voices and critically keep on searching for understanding what is at stake in the dialogue. This is a double-edged process of actively listening to aspects of community work, which might not always be straightforward, whilst at the same time critically examining the empirical material as well as one's own perceptions, not simply to reproduce the researchers' presumptions through the interpretation.

To do so my starting point for the interpretive work of this chapter is a whole range of different types of material. First and foremost, it builds on the documented collaborative work produced between the participants in the project: discussions, films, drawings, interviews, written notes and so on. Second, it builds on my own ethnographic notes throughout my fieldwork: conversations, experiences, reflections and ideas which have emerged in various ways through the work. Third, it builds on a wide number of semi-structured evaluation interviews undertaken with the project participants in order for them to provide their view on and perception of the project.[1] On the basis of these interviews, an explorative, critical hermeneutical analysis was produced in order to provide a broad picture of participant perspectives on the project. It is by constantly triangulating these various kinds of experience that the analysis of this chapter is produced. The text is my interpretation of the community work. The quality of the analysis is no higher than the level of analytic skills modestly combined with respectfully trying to understand the community perspectives at stake. The validity of the work first and foremost builds on the thoroughness of continuously testing any emerging interpretation against the various aspects of the fieldwork experiences.

This chapter in particular builds on the first half of the SuScit project; the community work and film-making projects, where the primary focus was for the residents themselves exploring and sharing their perceptions of living in the local neighbourhood. (An analysis with a more specific focus on the second part of the SuScit project, and the interactions between residents and researchers, will follow in Chapter 7.) Thus I am particularly building on the following parts of the community work: 1) my experiences with setting up the project in Mildmay, collaborating with the local community centre and other gatekeepers and recruiting residents for the three SuScit panels: the Young Peoples' Panel, the Lone Women's Panel, and the Older Peoples' Panel; 2) the resident introductory meetings and focus groups. These transcribed meetings held separately with the three resident panels were facilitated to openly explore perceptions of the local area and understandings of sustainability issues; 3) the community filming projects made in the three residents' panels separately. Through weekly meetings over a period of two months, residents discussed and were trained to put in pictures what they saw as central aspects to address related to the question 'what it is like to live in the local community'; 4) the First Shared Workshop, where all the residents, practitioners and researchers met for the first time. At the workshop we presented the community films as the starting point for discussing the themes addressed by the community participants, and for the researchers and practitioners to listen and learn from these experiences; 5) the Second Shared Workshop, where the participants through various visioning exercises we asked, on the basis of the residents' perspectives, to engage in a dialogue about what ideal futures of Mildmay 2028 might look like; and finally, 6) various others input from throughout the project with relevance for the resident perspectives at stake.

1 In total 26 semi-structured approximately one-hour evaluation interviews were produced with a broad range of participants across the various panels in order to explore their perceptions and evaluation of the project.

Garden and Grow

The idea of community gardening was not brought into the local area by the SuScit project – in fact it existed long before the project was made – but it was brought on to a common agenda through the project. Community gardening activities were one of the first issues which were presented to me by the local residents once I started working in the neighbourhood. The theme was brought up by a number of the lone women for whom the project provided a very welcome opportunity to actually meet their neighbours, talk about common concerns and share information helpful in everyday life.

Figure 6.1 Mayville Community Gardens

Source: Eames et al. 2009. Reprinted from the Citizen Science for Sustainability project.

Next to the Mayville Community Centre, between the social housing blocks, was a small fenced square of land, which local residents had got the opportunity to use as a community garden. A smaller group of women were gardening and had wooden boxes supplying their families with a few herbs and vegetables. Whilst the garden activity in itself could easily be elided by anyone visiting the area, through the community work and film-making it became visible how the gardening activities were not merely a question of planting vegetables, but key to these local residents in facing the challenges of living in the local area. In this sense, the garden was a lot more than merely a green square of land; it also has an important social function for the women. One of the women reminded me, while talking about the struggles of her daily life: 'You know sometimes I just got depressed, and then I am thinking that growing flowers must make people happy, so I am just trying' (Empirical research p12-LP10).[2] The garden is not least a free space, somewhere where it is possible to do something positive, as a counter movement to the daily conditions and concerns often perceived as beyond one's own influence. One of the activities in the community garden is to teach local children how to grow vegetables. Such activities have many purposes; besides growing a bit of food the women also see this as a pre-school qualification where children can learn and be better equipped to go to school. The garden activities are an opportunity to help bring up the kids. This aspect of resident-driven community work was not completely uncommon in the area: an elderly man organising an Asian lunch club explained to me:

> Sometimes, in the past I used to beg the young people who are free on Saturdays to help the elderly and that way we could teach them how to behave and how to deal with old people, it's an interaction between the old and the young, so that way they could get some experience and it could change their behaviour. So I think we are part in that. And when they come here we talk to them, and the old people talk to them, and in that way they get some sort of information about life itself. (Empirical research p12-OP5)

The community activities are an opportunity for the residents to make a difference to the area in which they live. Therefore the activities are not merely activities in their own right, but also 'answers' to the challenges the residents' face in their daily life. Of course the gardening activities have a value in themselves (for example the enjoyment of doing the gardening), but the activity also gets its meaning in relation to wider aspects of the women's lived lives. This was what began to unfold through the community film-making, where a group of women decided to make the film *Garden and Grow*. In the first part of the film one of the women shows the different colourful vegetables grown in the garden, noting that there is only limited space for doing so. The second part of the film concerns recycling, the production of waste and the dependency of an increasing number of Tesco supermarkets:

2 The system for referring to empirical research is explained at the end of the reference list.

> A lot of people are against this. These shops are conveniently placed near council housing but the produce is considerably more expensive than the local independent grocers or the larger supermarkets. They sell a lot of junk food and pre-prepared and heavily packaged foods which create so much waste. (Empirical research p04-LP)

The spoken message is delivered in the film while showing nearby graffiti providing critical comments on the so-called 'Tescopoly' of the city. The film ends with the following conclusion: 'We need more access to shops selling healthy foods at reasonable prices and ideally more opportunities for people to get allotments' (Empirical research p04-LP). In this sense the gardens are also a possibility for a bit of integrity and to provide some healthy food on a low budget, and thereby be less dependent on poor quality products offered in the cheaper supermarket chains.

In the Second Shared Workshop these themes were developed further. Through the visioning exercises, posters of visionary news of 2028 were developed in the format of newspaper front pages. One heading goes: 'Community cheers as Tesco falls' and concerns the following story: '800 people cheered and waved green flags as the last Tesco store was demolished by a wrecking ball to make way for small shops and allotments in the car park. Apple trees will be planted round the edge. A well provides water for the plants' (Empirical research p06). Two drawings show the former car park changed into green allotments. The newspaper tells the story of the local people reclaiming their area and being able to grow their own food rather than being dependent on the supermarket.

Another group at the workshop developed a related theme while making another newspaper article entitled 'Veg. surplus feeds London'. The story goes: 'Window box scheme huge success. Fresh vegetables and herbs for everyone in new tower block project leads to zero waste. Composting means no collections needed' (Empirical research p06). A picture shows a tower block full of green window boxes. The illustration contains many layers. On a very concrete level it represents a suggestion for how people in council housing practically could be able to grow their own vegetables. But at a more abstract level it can also be seen as the idea of people themselves being able to provide themselves with good food for living and being less dependent on the supermarket system. The story also notes that this solution produces no waste. Whether this would in reality be possible of course is an important question, but not the main issue in the visioning exercise. Rather, the message of the newspaper stories seems to be quite clear: *we want to be able to sustain ourselves*. In this way the newspapers represent a very clear resident response to the question of what sustainability means.

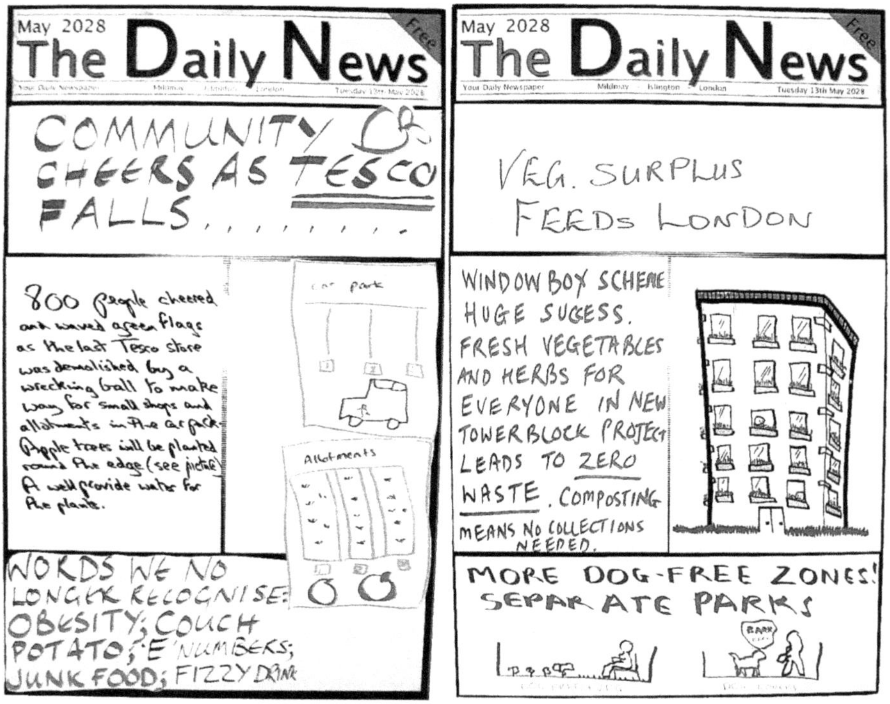

Figure 6.2 Mildmay 2028 visioning exercise

Source: Eames et al. 2009. Reprinted from the Citizen Science for Sustainability project.

The newspaper stories both represent a local community voice (about gardening projects in a social housing estate) but also a critical societal voice. The latter was rather clearly expressed by one of the women:

> "I think it's living in a city, it's a challenge". "[W]e have got caught up in this big cycle of living to work, and being caught in this kind of hamster-wheel of having to do all this stuff just in order to survive. And we don't have the time to socialise, to be in the area, to do nice things, so we're so busy just trying to survive". (Empirical research p12-L7)

The question of being busy just trying to survive is not metaphorical. It is a substantial challenge confronting the women in their everyday reality. To many of these people life appears as a daily struggle: the severe lack of money, time, security, a place to live, childcare, somebody who cares and so on were substantial issues of life, and for many the state of depression and lack of alternatives were certainly highly painful. A clear message on this issue was sent by the women's film *'Back to life. Back to reality'* picturing what family life is like in the deprived neighbourhoods of London: a struggle to survive. Community projects, and the social space between the women taking part in the project, offered an opportunity to cope with these challenges; 'to bring people together and break the barriers' (Empirical research p12-L10). A free space, where people through shared action, can claim a bit of integrity in life.

Figure 6.3 Mayville Community Gardens and the Community Centre
Source: Eames et al. 2009. Reprinted from the Citizen Science for Sustainability project.

Caught Red-handed

The first planned evening meetings with the SuScit Young People's Panel were postponed at the last minute due to youth conflicts between neighbourhoods. A young boy was taken as a 'hostage' and had his face cut by a number of older boys from a nearby area. Youth crime was a very real concern to the young people, and launching a project on 'sustainability' sometimes seemed somewhat naive in the tense atmosphere of these actions of youth violence. For many of the young people this theme was, for good reason, not their main interest or need. Rather it seemed that most of the young people turned up either for the incentive of money or with an interest in the filming activities offered. This obviously meant that the theme of sustainability was not what was really at stake for most of the young participants. While this in some ways was a challenge to the project, it was also an opportunity to listen to voices other than the usual suspects.

At the introductory focus group meeting the experiences of crime were evidently echoed. One of the key ideas which were brought up was the need for 'Memory boards for friends that have passed already' (Empirical research p03-YP): a local memorial wall for friends lost as victims of crime. Somewhere where the sense of loss could be collectively addressed and be made *visible*, and thereby *acknowledged*.[3] In this particular youth group two young friends were now only memories. Through the making of the film a group of boys continued working around this issue, producing the film *Caught Red-handed* (Empirical research p04-YP). The film starts by showing a boy walking through the neighbourhood passing a nice looking moped parked in an empty street. The boy recognises the moped, takes an extra a look at it, and drags it around a corner. As we watch the boy trying to hot-wire the moped, a group of boys, including the owner, catch sight of him. The group catch the boy and beats him with the crash helmet until he blacks out. They carry him to a basement room, leave him and lock the door. In the next scene we see the face of the boy waking up in the basement, realising the situation he is in. After a while the group of boys enters the basement room and collectively start kicking him from all directions while he lies on the floor. In the fight the boy catches sight of a screwdriver, grabs it and stabs one of the attackers heavily in the stomach. The final scenes show the boy escaping from the group, running out of the basement, leaving the building. The film ends with a close-up of the boy's face, running away.

During the making of the film one of the boys explained to me the importance of the last scene: the feeling of running away, but nowhere to go. In this sense the title of the film, *Caught Red-handed*, is double-edged: in a concrete sense it is an expression of being caught doing something illegal. But 'caught' can also be interpreted in another way – to be trapped in conditions from which you cannot escape. Originally the film was not meant to be about youth violence. Originally the idea was to picture the experience of conflicts with the police. However, as it was practically not possible to borrow police outfits, the script was slightly changed and the boy in the film was caught red-handed, not by police, but by another group of boys. While this obviously changed the content, the crucial idea of the film catches the main message:

3 The articulation of a need of a memorial wall can, building on the theoretical perspective provided by Shiva (Chapter 5), be seen as insisting that this issue is not merely individual, but in fact what can be seen as *a social common*.

Figure 6.4 Visioning exercise. Collage by the Young People's Panel

Source: Eames et al. 2009. Reprinted from the Citizen Science for Sustainability project.

getting into trouble; running away; but nowhere to go. The film does not merely show concrete actions, it also pictures a perception of the conditions for young people growing up in the area: *Where should we go? Where to address these problems?* Most of the young people had been living in the neighbourhood their entire life. This was the reality. Thinking beyond the horizon of the present was often not possible. Compared to the lone women and older people the life experience of the young people was more limited, and in this sense the statements of the young people were more representations of the-world-as-we-see-it-now than thinking beyond the present reality. These themes of crime were also heavily experienced by community workers in the area reminding me that the main challenge is for people to move out of poverty (Empirical research P5). For many people, however, physically moving out of the local area is economically not a possibility. Thus, as youth workers explain: either you get a job or you get into crime. That's the horizon, the two well-experienced choices of youth.

In the Second Shared Workshop these themes were brought up again in the visioning exercises, where the boys were asked to produce a poster on how they would like to see the future. The poster can be seen as consisting of three different elements: at the top a number of newspaper articles: 'They ripped out my heart and tore it into pieces' and 'Murdered Jimmy was our hero, say friends', which provide evidence of a common theme experienced by the boys. In the middle a number of centrally placed negations of the present: No Gun crime; No Gang Warfare; No Crack Heads. At the bottom a number of suggestions for a better future: More Football Pitches, More youth clubs, Safer Environment, Legalise Weed, Better Housing, More Green Spaces, More Schools. A common category for these suggestions is 'space' or 'places'. *We need some place to be.* Exactly the same theme can be found in the work of one of the other groups of young people. Under the title 'Space-Age', their visionary suggestion for the future includes: Arts facility; local shops; park; sports ground; communal washing facility; community centre (Empirical research p06-YP). In the dense local area this is evidently a substantial problem: there *are* no appropriate places for the young people to be. Further, the political response to youth violence has been making legal restrictions on young people gathering in the street. What is addressed in the project is the question: *Where should we go? There are no places to go.* One of the lone women commented on the present situation:

> When you hear what politicians have to say or whatever … they've got no idea really, they've got no clue, because they live literally on another planet. You know, they don't have to get on the 149 bus or the 73 bus where you know on these busses we've been threatened with knives, you know with (…) anything you can imagine, just on our local busses. And the politicians and people in power have no idea of what life's really like you know … travelling around in their chauffeured cars and whatever. (Empirical research p12-L4)

On the other hand it is also clear from the community dialogue that, despite not knowing where to address the problems, the community itself does not have the capacity to solve all the challenges:

> You see 2 year olds out in the street in the dark at night, you know what's all that about? You know, kids are not being brought up properly, and (…) kids I have seen have now been growing up and are in gangs, and you can see it right from 2 years old that it's happening, why isn't something being done about this from an early stage? Why isn't somebody saying "this kid is not being brought up properly" but nobody seems to care or do anything really (…) There's so many things you see when you live here but what power do I have to do anything about it? I'm hardly going to go to their parents and say "Do you think this is alright, you ought to bring up your child", you know. I'm likely to be punched in

> my face, probably, if their parents are at home. You know, I have kids knocking on my door saying "Can we come in and play" and I say "But we don't really know you and your parents don't know me". And they say "Well, we are not allowed home until seven o'clock and it's cold outside you know". Kids that are not allowed at home, you know. But it's like everybody seems to accept that this is just the norm ... there is something wrong there I think. So I don't know what the solution is. (Empirical research p12-L4)

What a Wonderful World It Would Be

In the first phases of the SuScit project we sought to address the particular question: What is it like to live here? Although the broader framework of the project was to discuss sustainability, the agenda for the community filming projects was for the residents collaboratively to explore their experiences and perspectives on living your life in the local area. In some cases, however, what happened appeared to be much more than merely a realistic hands-on description of the present state of life in the neighbourhood. The SuScit Older Peoples' Panel from the very beginning differed from other community and everyday life discussions in the area: the group was formed around the specific task of discussing local issues in relation to sustainability; the members had signed up on the basis of curiosity and interest in this idea; many of them were already community group organisers themselves; and the SuScit panel thereby joined people from usually segregated groups into one shared discussion. The meetings were based on very openly addressing the question 'what is it like to live here', and the facilitation aimed to create a trustful social space for the elderly to bring to the table what they would like to discuss; to have time enough to mutually share what they had on their mind; and to build on having no specific aim or product that needs to be reached at each meeting. Out of this work emerged a whole number of different discussions, interviews and community films. Anyone who systematically goes through this material will quickly notice the broad range of very different perceptions and statements on the issues discussed. One example could be the following two statements on what it is like to live in the neighbourhood:

1: 'You shut the gate if you've got a gate on your door, and the windows are barred, and you get used to living like it (...). I feel quite safe in there, but sometimes you look out and you look as though you're in prison, because you're behind the bars you know' (Empirical research p04-OP).	2: 'We together as a community help one another. It's a multicultural community with different nationalities living in this road, and we help one another, whereas if one is not well, the older generation would turn round and go and help those who needs help' (Empirical research p04-OP).

These two quotes might appear contradictory at first sight. An immediate interpretation, which is obviously partly right, is that the two statements are given by two different persons each having their own perceptions of life in the local area. The question is whether this is also the only explanation of the difference between the two. When systematically going through the material another supplementary explanation appears: the statements provided by the elderly are not only expressing what living in the local community is like. Often they are also implicit addressing what a better life *could be*. In the above example a better future might be developed from negating quote 1 (that is, not any longer to live in a community where you feel you are behind bars in your own home). Accordingly one could ask: is quote 2 a real statement of what the reality is like; or rather a visionary negation of the present (and possibly quote 1) pointing towards what is perceived as a better way of living? Once one starts examining how to interpret such statements, it becomes clear that is rather difficult to distinguish between the two (not only for the author of this text, but possibly also for the persons who make the statements). Who can judge whether a given statement represents a person's perceptions of how things are; how that person would like them to be; or something in-between?

In my analysis the ambiguity of *in-between* is a key element of what was addressed by the older people's panel. When analysing the issues addressed by the elderly as something in-between, the-world-as-it-is and the-world-as-it-could-be, the ambiguities of everyday life become clear. These ambiguities provide an opportunity for understanding the elderly people's statements not just as assertions, but also as expressions of *wanting* something in this world. The interesting point is not which of the above quotes provides the most 'valid' representation of what it is like living in the local community. The really interesting thing is that between these two quotes, in the tension between them, something important is going on, namely how people relate to a common concern. In my analysis a very important common concern of the older people can be phrased as the issue: 'how we live together in the community'. An example of this theme is the film *3 Steps* (Empirical research p04-OP) which in ambiguous ways shows one of the dilemmas of everyday life in the local area. The film essentially tells the story of two strangers, who by accident meet at the bus stop, feeling a personal need to talk and make friends. The film puts in pictures the ambiguity of getting the act together: wanting somebody to talk to/not speaking to strangers; greeting each other/just staying silent; sharing what is on your mind/looking at your watch as if you are busy; start talking and make friends/staying in your own solitude.

The film puts in pictures the ambiguity in meeting other people, and thereby addresses a key challenge of living in the local community. As one of the older men more generally notes in one of the film interviews: 'What a wonderful world it would be, if people solved their problems in a loving manner instead of going to war' (Empirical research p12-O4). 'What a wonderful world *it would*

Figure 6.5 Scenes from the film *3 Steps* by the Older People's Panel

Source: Eames et al. 2009. Reprinted from the Citizen Science for Sustainability project.

be'. Not 'what a wonderful world *it is*'. The very same statement both includes a lifetime of experience of what the world *is like*, and a wish for what it *could be*. In other words not only lived knowledge, but also *hope* is embedded in these statements. In these everyday life statements the inherent aspects of hope potentially connects *what is* and *what could be*.

Many similar examples can be found as part of the community dialogue. In the following example another old man replies to the question 'What do you hope for?':

> Number one: Unity. Unity among the races. Unity among the races. Leave colour out of it. You and I are striving for a better London; a better England; a better place to live in; a better place to bring up the kids. That's how I look at it. Leave colour out of it. (Empirical research p04-OP)

Still, despite all of the enriching conversations in the Older Peoples' Panel, after the end of the project, the multicultural group of elders in the SuScit panels tended to fall back onto pre-existing community groups segregated by ethnicity. As one of the lone women involved explained to me, evaluating the project:

> You guys were like catalysts; you were there, and you brought us together, and we were talking. As soon as the group broke, you know, if you saw one in the street and said "hello", you won't say anything more than "hello". I have no idea, I can't understand that myself either. (…) it's nice that when you see them they say "hello", because often you just walk past them. But then, I suppose, there is another fear; that when you get to know people and their personal details, and then you know, something bad might happen, it is not always a good thing. Welcome in London, you know. (Empirical research p12-L10)

In some cases the community discussions of the project brought something to life, which seemed to rarely happen: the possibility for sharing *hopes*. The examples presented above do not just represent community life *as it is*. It represents a discussion of what everyday life is and what it *could be*. Thus, your reading of these words needs to have in mind that this is not merely descriptive research, it is action research, trying to open an arena for sharing views and perceptions of what a different future might be like. In the older peoples' discussions a key aspect was the sharing of what is valuable in life: 'my garden; light; my grandson crawling; good time with friends; delightful conversations' (Empirical research p04-OP). Insisting on addressing the *beauty in life* should not be taken as a statement that life is easy. Rather on the contrary, vulnerable hopes and dreams are formed by and in close relation to the rough realities of everyday life.

Figure 6.6 The Older People's Panel in the Mayville Community Garden

Source: Eames et al. 2009. Reprinted from the Citizen Science for Sustainability project.

Sustainability in Everyday Life

What I have tried to illuminate in the above examples is what I perceive as some of the key aspects of the community dialogue, which came into being through the project. In terms of showcasing the entire range of subjects and issues which were brought up, many more examples could be given. My particular interest, however, is not to provide a complete account of these, but rather to provide a starting point for grasping the nature, potentials and pitfalls of such a community-based approach to sustainability. In the following I shall show how I perceive and interpret the insights from the community experiences.

An initial observation is that enabling a social space for addressing what it is like to live in the local area might make room for articulating all sorts of experiences which are clearly perceived as *not sustainable* in everyday life. The examples above provide some good illustrations in this respect: *We need some space to be*; *We need a bit of integrity*; *We need love and friendship*. These are needs, which are not perceived as fulfilled, and which in a sustainability perspective can be interpreted as a very basic claim: *We want to be able to sustain our own lives*. At the same time it is evident that this is in fact a substantial struggle. In my interpretation, the examples show how states of social erosion seem to question fundamental self-sustaining conditions for human life. The message is double-edged: on the one hand, reclaiming what is perceived as important to sustain everyday life (for example the garden, places to be, the sense of community); on the other hand, addressing that the problems are unlikely to be solved, merely on the basis of personal and community-based capabilities. The question still persists: Where should we address the challenges? In this sense the issues exemplified above represent areas, where society is perceived as failing to solve a broad range of fundamental problems in modern urban life.

The three key examples of this chapter primarily concern what can be interpreted as *social* aspects of sustainability. This is not to say that environmental dimensions were not part of the community dialogue. But it is evident that 'the social' were given much more weight by the residents than 'the environmental', possibly because social erosion in this case appears to impact more directly on people's everyday life than the environmental. Accordingly one could ask: is there, within in the economically and socially deprived areas of a mega-city like London, where social challenges are fundamental aspects of everyday life, any concern about the environmental dimensions of sustainability? Despite the urgency by which social concerns were addressed, the notion of nature also emerged as a locally grounded feature of the dialogue. Parks, gardens and green places to be were articulated as an urban privilege *in* social life, as *part of* social life. The community garden provides a very clear example of this: the garden as nature is a social space, with different opportunities than other urban places. Thus, this notion of nature is directly interlinked and emerges from everyday life.

Another aspect of the community dialogue should be mentioned as well: Environmental concerns were not only local but also global. Perceptions of climate change as a serious challenge were clearly present within the community. As one of the older participants reminded me:

> Tell you what, this is a global issue! I know we talked about the local, but at the end of the day it is gonna be a world issue. The more people in the world, there will be no more water, less food … maybe not for your children but for their children's children, and in time where will be fights over food and water and everything. (Empirical research p12-OP5)

The global dimension were certainly present in the residents' minds, many of which had family and personal relations in countries where environmental damages where clearly more severe than in London. However, the discussion

> (…) it's kind of an important thing to reduce your carbon footprint, as again global warming is a major issue around the whole world, it's affecting everyone. So I think if everyone made a kind of small difference, then it would make a big difference, because there are six billion people in the world, so six billion differences makes like one huge difference. (Empirical research p04-YP)

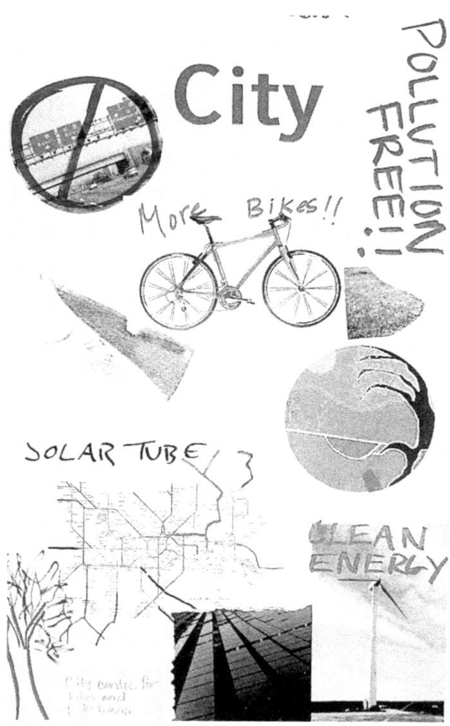

Figure 6.7 Environmental icons in the visioning exercise

Source: Eames et al. 2009. Reprinted from the Citizen Science for Sustainability project.

of climate change also seemed to add a more abstract level to the discussion, by drawing on expert-led discourses, for example, the notion of carbon footprints.

In my view the concept of carbon footprint represents an expert-led discourse echoed in a community voice, rather than something emerging from everyday life itself. Several examples of such mediation between expert and lay-people aspects of sustainability can be found in the project, for example in terms of 'environmental icons' such as windmills, solar-panels and so on added to the community perspectives. What was distinct about the community work, however, was that everyday life-based notions of sustainability were building on experienced aspects of life. These perspectives were not merely about how things are, but also how they could be. They inherently build on hopes although these were not always expressed directly (hence, they can be interpreted as what Nielsen and Nielsen characterise as 'manifestations of life').

It would be a false extrapolation to suggest that the community perspective simply shows the way ahead towards a sustainable future. Taking into account the challenges addressed in this chapter, furthering sustainability *is* a broad societal challenge. In particular it must be acknowledged that major environmental challenges today are being managed by complex expert systems, and that such systems of expertise accordingly have an indispensable role to play. What the community perspective offers is to root sustainability thinking in people's everyday lives by listening to people's identifications of what appears to be unsustainable in modern life, and aspirations of a better future as a potential for furthering sustainability. In the next chapter I shall look more into how that went on in the social meeting between local residents, practitioners and researchers taking part in the project. In doing so a particular interest is to understand what happens in the meeting between the ambiguity of everyday life and the dynamics of academic culture.

Chapter 7
Social Learning between Citizens and Scientists

The particular purpose of this chapter is to address the question whether, and in what sense, social learning took place between citizens and scientists taking part in the Citizen Science for Sustainability (SuScit) project. Reflecting the overall research interest of this book, particular attention is paid to understand the dynamics between researchers and residents, still acknowledging that the practitioners involved were an important part of the project.

To empirically understand the dynamics of social learning is crucial to this book. Moving public engagement upstream not least implies a movement from the battlefields of policy-making towards the epistemological fields of research and development (acknowledging that the two are interlinked as well). However, while 'downstream' engagement largely builds on raising citizen perspectives as specific positions on science and technology in societal political debates, the potential impact of 'upstream' engagement is much more dependent on epistemological processes of mutual learning. If upstream public engagement does not leave any epistemological footprint in the world of research, it is hard to argue for its legitimacy. Thus, analysing whether mutual learning takes place appears as an acid test of any project perceiving itself as upstream public engagement.

Whilst the analysis in Chapter 6 was highly hermeneutical, this chapter builds on the former by adding a more critical-analytical perspective: to test the SuScit project against a given set of criteria for social learning. In doing so I am building on the theoretical framework provided by Nielsen and Nielsen introduced in Chapter 5 (1999; 2006a; 2006b; 2007; 2015). My analytical strategy has been to identify a set of criteria which appear as key points in processes of social learning, and which I shall briefly outline in the following.

A first analytical criterion is to look at social learning as an epistemological process *between distance and engagement*. On the one hand, understanding is only possible by socially engaging with the other; on the other hand it also builds on distance and difference between peoples' perspectives. Oral articulations and representations thereof are merely one aspect of epistemological processes of social learning taking place between and within individuals.

A second analytical criterion is the question of grasping *the societal dimensions of particularities*. That is, being able to, on the basis of the particularity of everyday life issues, mutually share and explore possible societal dimensions perceived as part of the specific issue at stake. Thus everyday life issues become initial points for a reflexive process, on the one hand grounded in the particularity of everyday life experiences, while on the other not constricted to but transcending this horizon. In this sense social learning means mutually exploring connections between the particular and the societal.

A third analytical criterion is the potential emergence of *social imagination*: creative processes unfolding aspirations of life and based on lived experience start imagining alternative ways and concrete ideas for how one might like to live in the future. In this perspective the social learning perspective is not merely about establishing shared understanding, but questions to what extent the social process implies that participants more actively start relating to the issues brought to the table by developing a sense of obligation to act upon new insights.

The three criteria above should make it clear that social learning is a social process. The 'social' is not an add-on, it is a prerequisite: the media through which social learning takes place. Epistemologically speaking, the emergence of shared human understanding is a social process, which although often building on, cannot merely be reduced to linguistic representations. In Nielsen and Nielsen's perspective, linguistic representations (or other forms of mediation) are traces of that process: footprints in the social landscape pointing in the direction people have moved, but not the movement itself. In this respect my methodological approach builds on, on the one hand, hermeneutically to understand what was at stake as seen from the perspectives of the various participants, and, on the other hand, as a researcher playing an active part, critically interpreting these perceptions and 'testing' them against the focal points of social learning. In doing so I am building on the documented collaborative work produced between the participants (discussions, films, drawings, interviews, written notes and so on); my own ethnographic notes throughout my fieldwork (conversations, experiences, reflections and ideas which have emerged in various ways through the work); as well as the critical hermeneutical analysis of the semi-structured evaluation interviews undertaken with the project participants. It is by continuously triangulating these various kinds of experiences that the analysis of this chapter has been produced. The text is my interpretation of the SuScit process, and the validity of the work builds on the thoroughness of continuously triangulating any emerging interpretation against the various aspects of the fieldwork experiences. My research interest in doing so has not been to paint a full naturalistic descriptive picture of the project, but to understand in what sense aspects of social learning were at stake.

Whereas Chapter 6 in particular built on the community work of the first half of the SuScit project, this chapter develops by focusing on the shared engagement of the second part of the project; the various workshops and meetings whereby participants meet to collaboratively share, explore and build on their various experience, that is: 1) the First Shared Workshop where residents through the films and discussions shared their perceptions of what it is like to live in the local community, providing the researchers and practitioners the opportunities to listen and learn from these perspectives; 2) the Second Shared Workshop where all participants were invited to engage in shared visioning exercises of ideal futures in a 20-year perspective; 3) the Researcher and Practitioner Workshop where the professional participants worked on developing ideas and initiatives responding to the issues raised through the community dialogue; 4) the Third Shared Workshop where these ideas were responded to the local community; as well as 5) the various dissemination meetings, and further development of initiatives based on the SuScit collaboration.

Before undertaking the following analysis I would like to make clear: the SuScit project was not based on social learning theory. Accordingly the following analysis is not an evaluation of how well the project met a set of preset goals. The project was built on a development of best practices aimed at moving public participation further by fostering direct dialogue between researcher, practitioners and residents. As such, the project was similar to many other initiatives in the field of public engagement, which are first and foremost part of a field of practice, and, second, objects of academic studies. Why then, one might ask, is it relevant to undertake a social learning analysis of such a project? In my view, because it might further the understanding of crucial aspects of the project. From my direct involvement, as well as various evaluations and analysis of the project, it appears evident that the project actually fostered a lot more than the planned production of interlinked thematic discourses on sustainability. This can adequately be described as the creation of social space(s) across residents, practitioners and researchers. Within the framework of action research such social spaces are interesting points for analysis: What happens when new social spaces emerge?

Between Distance and Engagement

If the SuScit project enabled shared social spaces among the participants, how can we understand what that meant to those taking part, and to the thoughts, actions and reflections based here on? One first observation might be that the SuScit narrative, to gather the residents to articulate their perceptions of what it is like to live in the local area, and for researchers and practitioners to listen, learn and reflect on the issues brought up, obviously challenged more traditional perceptions of the role of experts. As one of the researcher notes:

> I found it difficult to sort of switch from the "interviewer"; you know when you normally act in a research mode and you are trying to tease out what people

> want and what they think (…) the idea is that you don't impose yourself on the subject that you are really trying to absorb and see things through their eyes. And I found that kind of exercise quite difficult 'cause in a way we were doing that and (…) it's coming back to the same point, about saying "we do have this expertise and this is what we can give back or comment on". (Empirical research p12-R5)

The reason that the interviewer role appeared difficult was not least the format of the workshops urging everybody to engage with other people. Engaging in a shared group exercise (with shared practical activities like the newspaper visioning exercises which build on shared collaboration) it seemed somewhat awkward as a researcher simply to interview people. On the other hand, most professionals were very aware that the setup was for them to listen and learn from the resident perspectives. Hence, the question appeared in a practical social setting: what does it epistemologically mean to listen? In the evaluation interviews one of the researchers recalls the experience of the first community discussions at the Shared Workshops, thinking, 'Oh gosh, if we left planning to these people, it would be totally environmentally unsustainable' (Empirical research p12-R2). But the researcher also explains how things continued. A particular example is a dialogue with an older man articulating his hope for the future: to have a huge house. The idea of starting to build huge houses in a dense urban area clearly falls outside academic categories for future urban sustainability, where space and resources are highly limited. However, the researcher continues the conversation with the man, asking *why* he wants such a big house. And the man starts his explanation, revealing that what appears at first sight as a material aspiration is in fact also the articulation of a social need. The researcher reflects on the example:

> I think, what we need to do is to get behind, what it is that those spaces are giving them. (…) 'cause actually one guy wanted a really big house with six bedrooms. He wanted his whole family in there, so he'd have, I suppose, grandparents, parents and children. So it's not about two people having a really big house. So what that is saying is being able to keep the family together in the community. (…) [I]t's actually taking: "Ok, this is physically what you said you would like", but it's working out, what is it behind that you want (…) And actually what people says is sometimes just a shorthand for saying other things. And it's just that conversation I had about: "Ok, why do you want six bedrooms and a house on its own?" And then he started talking about: "Well, because with my children, my grand children". And then you think, oh well, that's a different kind of … and it's giving people the space to develop those ideas. And in some way SuScit was such a luxury, in a way, because you had all those different meetings and people explored different ways. (Empirical research p12-R2)

The example has several interesting aspects when looking at it from a social learning perspective. Let me try to illuminate these by addressing the question of where the 'added value' emerges in the above dialogue. Obviously the capabilities of the researcher, critically asking for and making analytic interpretations of the issue, are crucial. The conversation *could* have ended with the researcher concluding 'this is far from sustainable', but the dialogue continues towards a new insight, both building on the everyday life aspirations of the older man, and the knowledge and analytic capacity of the researcher, concluding that we need to get 'behind that you want'. However, academic capabilities alone do not explain the above. The dialogue also builds on a social process in which the man trusts that he can share his life aspirations, assuming that the researcher listens to him. This implies that, within the workshop setting, there appears to be some kind of social space for people to share hopes and aspirations, which obviously can be a vulnerable thing to do. Analysing how the participants themselves reflect on this, it is evident that a crucial part of this process was a personal engagement and commitment to a common process. For the residents the difference between hope and despair can be rather small, as articulated by a woman reflecting on the question of why she took part in the project:

> I don't know.
>
> It was, like I said, a dream.
>
> And everybody has to have a dream.
>
> And if they don't wanna listen, that's ok.
>
> You just wanna show them your views,
>
> And that's why I did it. (Empirical research p12-LP10)

The hope that somebody wanted to listen, and that things could be changed for the better, was clearly an aspiration for many community participants. But the commitment was mirrored by the researchers and practitioners taking part, which became clear in the later phases of the project, where the professionals were asked to reflect on the community issues raised, in order to develop ideas for what would be appropriate responses addressing these issues. In the evaluation of the project, a common theme for the researchers and practitioners is the paradox of, on the one hand, being able to articulate numerous important new initiatives supporting the local community, whilst, on the other, for various reasons, not being able to find sufficient support to establish these initiatives. The dilemma illustrates that taking part in the project was not merely a question of gaining new insight; it also became a question of engagement and commitment amongst those involved (and one could note: exactly this dilemma was an often experienced paradox of much of the research; in fact many of the researchers involved hoped that SuScit could be a way to overcome this distinction).

Taking into account that social engagement was in fact a crucial aspect of the process, what role did it play? One of the researchers explains this issue in greater depth:

> I really enjoyed it because I found people who were of my demographic; women my age living in the same part of London; with similar kind of experiences; and similar kind of interests. And we could kind of swap information about things going on. In that sense that was kind of easier than I had expected – whereas I didn't really have much interaction with the youth or the elders. I choose to go with the women because I had something in common with them and I felt comfortable with them. I had … a shared experience to start from. (…) It takes longer to build up a kind of a pool whereby you can engage with others on a level of comfort. It was possible to happen much more quickly between me and them. I don't think in that level of timeframe I would have made the same level of pool with the youth or with the elders. Before you can get to a point where you are able to listen to people and to communicate with them, you do need to have some level of a pool. And I think this is a problem with researchers and local communities or professionals and local communities that things are often done in a very short time frame. I think that's a problem, that it does just take time, and what was good about this project was that it was over a reasonable time, and there were a number of chances for people to engage, and I think there was a reasonable diversity of people. (Empirical research p12-R1)

In the above case, understanding other people is not just a question of recognising the words coming out of their mouth. Understanding other people is much easier if you can draw on some level of resonance in your own pool of experience. Thereby social engagement becomes a shortcut for understanding other people. In other cases the lack of a common 'social pool' could be a hindrance for sharing and understanding people's perspectives on the world. This links to a rather basic challenge; how to share human life experience? One of the residents notes:

> Some of the professionals could do with perhaps living in the real world, well this reality, to experience first-hand how things are. But then … they experience a different life. (…) perhaps they have one intention and everything, but they just can't put their head into that space of being somewhere like here, they just don't understand. (Empirical research p12-LP4)

The reality of living in a deprived neighbourhood of London can be communicated in various ways. But does that mean that people without that first-hand lived experience can actually understand? Obviously the potentials of shared understanding have their epistemological limitations. However, it should be noted that something happens when bringing researchers, practitioners and residents together using methods aimed at establishing more equal forms of dialogue. In that social space professionals can no longer argue merely by referring to abstract explanations. They might have one in the backs of their heads, but in the social conversation, lay-people and experts must be able to meet face to face in order to establish some level of two-way communication. For many professionals this was far from easy: What to say, if you can't refer to privileged expertise? What happens in this setup when academic knowledge gets confronted with the harsh realities of everyday life? In the researchers' evaluation of the project a common reflection is the question whether residents are perceived as empirical objects or part of a more mutual interaction:

> You know we have these ideas (...) we think about these things, we talk about them, we write about them, we read about them. But its having them sort of challenged I suppose, when you discus them with local residents and what they think is important, and how they are defining sustainability. It's listening I suppose and learning from that. And what I'm saying is one step in getting a creative dialogue as well as a result of that, so that there are change on both sides, potentially change on both sides. (Empirical research p12-R7)

Social engagement is not merely an intellectual exercise, it challenges the role of being a researcher; it disturbs academic thinking. And one might ask: are such interventions appropriate; are the articulated voices of the community valid in research; is it taking research forward or backward? Such critical questions obviously must be addressed in any participative setup. Here I shall merely try to unfold what such participative approach might imply for the researchers involved.

> I think, being engaged with real people, with real needs and desires, I think motivates you to do better work, and more timely really. You gonna get your act together if somebody is actually waiting on you. It makes it more meaningful all round. And then hopefully you have given them something good as well. (Empirical research p12-R9)

For the researchers, what grew out of the project was more than sharing various forms of knowledge; it also became a matter of sharing hopes and motivations, thereby addressing the question *why are we doing research in the first place?* Obviously in terms of the actual research practice, answering that question remains the privilege of the researcher. However the dialogue seemed to provide a broader societal context of understanding and readdressing this question, through what can be seen as an epistemological movement between distance to and engagement with the community:

> It's got me thinking about the kind of research I do. [A]t least [it] makes me think wait a second, I'm doing research that is supposed to be benefiting the sustainability of cities, then actually talking to people, but in a way that I'm listening to them and trying to really take on board their ideas, and not trying to fit them into a prescribed sort of theory or notion or ideas, but really working with them. I think it helps me more. (…) And then just on a personal level, less professional level, it's just thinking about (…) how I'd wanna create a better community for where I stay. (Empirical research p12-R12)

Thus, we are getting closer to a crucial aspect of social learning: social learning is not merely about understanding other people's perspectives. It is essentially about actively relating to these, in order to develop shared understandings which have importance for those taking part. If that happens, a key criteria for social learning is not merely whether new insight is created, but to which degree the social process calls for or encourages a sense of obligation to act upon the insights collaboratively developed through the process. Obviously this can happen at many levels, ranging from hands-on community responses; researchers changing perceptions of the work they do; and strategies for change at higher levels. In the following I shall further explore in what sense this took place in the SuScit project.

Societal Dimensions of Particularities

In the previous chapter I sought to show that the particular issues raised by the residents were not merely concrete questions, but also embedded in and thus part of wider societal contexts, being addressed in particular ways. This was the case for the young people, the lone women and the elders raising rather basic needs in human life. *We need a bit of integrity. We need some space to be. We need love and friendship.* These examples were not merely local particularities, but societal challenges, perceived as currently not sufficiently acknowledged by society. In this sense the issues discussed at the shared workshop were potentially, although locally grounded, also points of departure for wider societal questions.

One of the interesting aspects of the way the SuScit project was designed was that it placed community voices centre stage, as the starting point for a shared process among residents, practitioners and researchers. Thus the issues and themes raised by the residents themselves provided a framework for the wider discussions confronting academic concepts of sustainability with a whole number of community issues, each addressing very concrete challenges and more general societal questions. Through shared reflections over the community work a societal dimension was brought to the project. This feature of the project was subject to quite a lot of reflections amongst the researchers. 'All research is done in a very local context, in the laboratories down here with a particular kind of interest and issues, but (…) the main kind of output for academic research is something that is more universal, more global than particular' (Empirical research p12-R1). 'Global' in this sense does not refer to the term societal, but to the global academic community. Thus the local dimension of the community work confronts academic research interests. A researcher comments:

> Academics, who are generally middle class and many other kinds of particular demographic indicators, living in that kind of global class and don't really (…) in any kind of local community. (…) [T]he people that do the research tend to be, you know, relatively affluent, come from all sorts of different parts of the world and are recognised for having outputs again on that kind of global stage. So their accountability is in no way (…) in any kind of local community UK context. (Empirical research p12-R1)

This distinction between local and global was a critical issue to many of the researchers finding that current academic incentive structures further academic accountability to global research societies, rather than active involvement with societal contexts at various levels. The legitimacy of research is first and foremost the academic novelty of invented ideas, not their usefulness in a societal context. In contrast, to most of the practitioners it appeared as an inherent paradox, that the main aim of the SuScit project, to deliver inputs for an overall research agenda, became a key barrier for focusing on delivering some real-world tangible outputs. Thus, the triangulation of the three groups of researchers, practitioners and residents at least made clear that defining societal challenges of sustainability was not a privileged task of academics.

A more adequate understanding of the role of researchers in the project probably was to translate local and societal issues into academic ways of coping with these. In order to get a deeper understanding of how this 'translation' took place, it is useful to have a closer look at the discussions which emerged on the basis of the three cases already brought up: a) the women's community gardens; b) the issues of youth crime and lack of places to be; and c) the community perspectives raised by the elderly people.

The issue of community gardening brought up by the lone women fostered a number of initiatives in terms of financial support (to develop the local gardening project) and institutional strategic response (in terms of collaboration with the municipality in relation to approaches to sustainability and community engagement), as well a whole range of different proposals for research and further action: critical scientific analysis (a research proposal for analysing whether local food production is environmentally sustainable); development of applied technical solutions (ideas for education projects inviting engineering students to work with local residents developing new applications of technologies based on needs identified by the community); and research for social change (action research proposal aiming to counter the effects of gentrification through social activities of community food growing and production) (Empirical research p07;p08;p09). The example shows various ways in which issues of community gardening were translated into separate subsystems, each having a distinct approach of responding to challenges faced by the local community. The example on the one hand shows that various kinds of 'system responses' can be developed on the basis of issues articulated locally. The suggested initiatives are responsive to the local challenges, in the sense that they are responding on the basis of the particular opportunities provided by the particular subsystems. But on the other hand the examples also showcased a gap between clear and outspoken aspirations of researchers to further community engagement, and the actual difficulties in establishing initiatives building on new orientations in collaboration between researchers, practitioners and the local community.

A further discussion of the potentials for academic research cultures to actively build on community engagement is an issue I shall leave for the next chapter. For now let me continue by discussing a few more aspects to the 'processes of translation' by drawing on the issues about spaces brought up by the young people. Amongst most professionals, many of which were struck by the brutal reality of the film *Caught Red-handed*, there was a broad consensus that crime-related issues were among the more important issues to the local area. However, the actual uptake and impact, as part of the practitioners and researchers developing ideas for research and initiatives responding to the issues raised, was more limited. The professional participants themselves noted this problem during the Researcher and Practitioner Workshop:

> It's a bit different to separate out what's most important from what's most interesting (...) I didn't get so excited about crime, but that's not because it's no less important, it's not just such an interest for me. (Empirical research p12-R2)

In other words, for research to take up challenges brought about through community engagement you need the involvement from specific research fields able to perceive the addressed challenge as a research issue. On a practical level one might add that this is an important task in the planning and facilitation of a public engagement process – to ensure the right people are on board to foster synergies between issues at stake. But at a deeper level this is a substantial challenge: What if the community voice falls between two stools? Addressing the question is important because it is well known that this is actually a key feature of everyday life knowledge: it does *not* simply fit into pre-determined categories. One of the research ideas, addressing the issue of lack of space, was the formulation of the following question produced at the Researcher and Practitioner Workshop: 'How can different experiences (youth, woman, elders) and perspectives be incorporated into urban design to reduce crime?' (Empirical research p07). On the one hand the question acknowledges that there is a whole range of local experience, which could more generally be taken into consideration in new research projects. On the other hand, the question also stands in contrast to the young people already providing a rather relevant answer: we need places to be. The example illustrates that what is possible to bring into a research agenda is slightly different from the perception of pressing issues on the community level. Obviously the professionals involved in the project qua their employment all had a privileged position in addressing issues on the societal agenda. On the other hand the space for defining such issues is obviously framed by the institutional foundation on which researchers and practitioners operate. One of the researchers provides the following reflection on the process of community engagement:

> One of the things that sort of always struck me about environmental attitudes and environmental policy is that the public is usually way ahead of the politicians in terms of what they would be ready to support in terms of changes. So what's interesting here is that it goes through the population. (Empirical research p12-R6S)

What happened to the issue brought up by the older residents on community coherence, love and friendship? Some of the professional participants argued at the Researcher and Practitioner Workshop that this dimension should be part of *all* specific parts of the project. In many cases traces of this dimension can be found in the discussions among researchers and practitioners. But at the same time it must be noted that it proved rather difficult to fit this community dimension into academic categories relating to sustainability. This difficulty deeply confronts researcher and practitioner approaches to community engagement. As noted by several of the professional participants, furthering community cohesion is a question of working *with* the community, avoiding treating people as a research lab. However it is rather clear that while community cohesion was perceived as an end by the residents, for researchers it also became a means for furthering other aims (such as sustainability agendas), thereby highlighting that many different interests are at stake in community research. One of the residents provides the following evaluation of his involvement in the project:

> I wanted to hear: is there anything humans can do? And I came to that conclusion: they cannot. (…) Because of all that good talking; nice ideas; fluent speakers; they know how to put things in words, so it sounds nice … but when I look back in history, all the eloquent speakers, what are they able to do really? I know I sound negative, but that's the reality. (…) those concerns: environment, crime, injustice. Nobody has been able to solve that, no government or human can solve that problem (…) Humans can do changes about the material things, make better fridges and better homes, but when it comes to the basics they cannot. (Empirical research p12-OP4)

Hence, the question emerges: What are the horizons for this type of engagement between researchers, practitioners and residents?

Towards Social Imagination

If upstream public engagement has something to offer, it must at least leave some kind of epistemological footprint. However it is clear that such a development is hard to evaluate. The SuScit project was both an alternative initiative trying to challenge the academic agenda of sustainability, but also inherently embedded in specific historical, institutional and societal contexts. Accordingly I find it hard to show that the project 'itself' caused this or that development, simply because the project is not completely separable from its context. Instead I find it more useful to analyse what tensions emerged through the project, because such tensions show that something was at stake. In this perspective I shall address the question in what sense social imagination emerged, recalling the definition provided by Nielsen and Nielsen in Chapter 5: 'Social imagination is related to the gesture of performed and unfolded *life aspirations*, concrete ideas about, how one could live (…). The formulation of potential life aspirations are characterised by *pre-empting a different – a better or fuller – life*' (Nielsen and Nielsen 2007: 31–2).

Examining the community work establishing the first half of the project, several examples can be found which can adequately be understood as aspects of social imagination emerging amongst the residents. The theme 'what a wonderful world *it would be*' (see Chapter 6) catches the *Older Peoples' Panels* discussions of community cohesion, both in terms of describing social difficulties of the present, and as aspirations for a better different future. In this sense, this part of the community work provided a social space, *a free space*, for collectively sharing aspirations and thinking differently beyond the present reality. This is an essential part of the concept of social imagination: the emergence of social imaginative processes transcending the present while still being rooted in experiences of lived life. Examining the *Lone Parents' Panel* in this perspective, the women's work with the community garden can be understood as a social and physical manifestation of social imagination: the garden initiative can be seen as a 'counter answer' to problems and struggles experienced in modern urban life, mobilising a self-empowering process of social activities. In this sense the garden is a realisation of imagining alternatives aiming to overcome some of the aspects of unsustainability experienced by these women. The *Young Peoples' Panel* had in much of their work a strong orientation towards negating unpleasant aspects of the present. The idea of imagining alternative futures in some ways proved more difficult, as was it to think that one could have an impact changing the conditions of the present. Why should one share hopes and aspirations if not having the experience that it might make a difference? That being said, many of the young people had strong ideas about how the future needed to move in other directions. But the difficulties in the processes underlined that social imagination is not something that can be forced through, it is something that might emerge if the conditions are right. This in fact is an important point understanding the concept of social learning outlined in this book: social learning is not just a concept emphasising that most learning processes take place amongst other people. It is a question of the emergence of *sociality*, whereby societal aspects of particular experiences in lived everyday life become visible, and where people, by sharing aspirations about how they would like to live, start imagining alternative futures, and ideas about how they might come into reality. The strength of the community work was that through the discussions and films, aspirations and ambiguities of everyday life in the local area become visible and shared.

What happened then in the *second* part of the project in the dialogues between citizens, practitioners and researchers? In analytical perspective at least two different processes were going on. The first can be understood as a mediation of different understandings of sustainability. On the one hand, the visioning exercise made it possible to articulate aspirations for future alternatives to the present, both in terms of identifying challenges of the present and visions for what a different future could build on. But on the other hand, the workshop dialogues also became mediations between different perceptions of what sustainability is. This was the case when different concepts and understandings were confronted between residents, practitioners and researchers. But the tension between everyday life experience and expert concepts of sustainability was not least mediated by residents themselves. In contrast to the community film-making, all sorts of 'icons for sustainability' were added to the community visions by the residents: solar panels; windmills; bikes; no-carbon-emission cars and so on. When comparing the resident work in the first and last part of the project, many of the key messages are the same, but the means of communication are different, they are much more targeted at a 'sustainability audience'. In this way citizens' aspirations of the first part of the community work was translated into perceptions of 'an academic language of sustainability'. Hence, one could ask: is that to be termed 'social imagination'? My analysis is that social imagination in the project was first and foremost something that emerged in the community part of the project, and was brought into the shared deliberation with the professionals, where ideas were mediated between many different understandings amongst the different groups of participants. Through this process, researchers and practitioners were 'translating' community perspectives into something which could both make sense in their own professional area and at the same time appear as an adequate response to the community. In this way the community perspective broadened the context of sustainability research, which could not merely refer to arguments counted as valid by academic communities, but also had to take into account the issues addressed by the local residents.

Such a kind of process, however, is not necessarily in line with the theoretical concept of social imagination. In fact, when not concluding that social imagination, in its theoretical sense, was not at stake in the shared workshop format, I build on the observation that for many of the professionals, meeting the community perspectives called for reflection on the role of science, how it is being done, and whether it might meet the challenges and aspirations addressed by the local community. In this way the community perspectives called for reflexivity and thinking differently about academic approaches to urban sustainability. Understanding this process analytically, the shared workshop became a hybrid between building on moments of social imagination and mediating pre-existing discourses of sustainability. It is exactly in this diverging understanding of what is understood by 'social learning' that the empirical tensions in the process can be found. These tensions not least became visible through the evaluations of the project. One of the researchers notes:

> I don't think involving the public will ever really going to come up with the innovative ideas we need to. And I think that is quite a difficult one for me to grapple with, 'cause I really think it's important to have that. But I think the kind of ideas that we need, will come from individuals with just that ability to think outside the box. Sometimes they do come from the community, 'cause sometimes you have really very unusual people. (…) when you get really powerful change in a community what is it that creates that? (…) you have the right political economic situation, then you have a few key individuals who have the vision to identify what can be done (…). I think it is a tension, because unless the whole community owns something, then the changes you need to make aren't going to work, but on the other hand (…) I think it's too much to expect the community to come up with ideas all the time. So I think there's a kind of balance and (…) I think there is a limit to how much public participation can deliver. (…) you need to be able to nurture the people who are going to have that break of vision and

> the ability to get it done. But you also need the community to own the changes to be made. (Empirical research p12-R2)

These considerations certainly link back to understandings of the role of researchers. Another researcher put it this way:

> There are a lot of different roles: The *listening role*, and in fact as a researcher you are there to find out, and sort of as I say enable and facilitate people to explore issues in more depth (…). But then there's the other role, I suppose, because researchers do have this more kind of *strategic role*, they got information from a range of different things, so it's sort of feed in information about what's going on in different places and different ideas, which is one role that researchers can take. And also putting the local experience in that more strategic kind of context. So making the links, making the connection between what's going on locally, both in terms of policy agendas, you know, what can we learn from elsewhere and what things can be brought in there, but also some of the ways of thinking about things as well. (Empirical research p12-R5)

If what can be learned from the SuScit project is what tensions arose during the project, *questioning the horizons of doing research* was clearly brought to the table. In this sense what can be seen as social imagination was not so much about novel research ideas or ground-breaking community initiatives; rather the focal point seemed to be questioning research itself and the way research is being done:

> Providing more kinds of opportunities to actually get to know what life is like for people who aren't like me; I think it's really important for sustainability researchers, 'cause it's far too easy to just think about communities and the public or citizens or whatever, in terms of our own kind of experience, which is very different of someone from the Mildmay Domino Club, you know. And the people who are in the Mildmay Domino Club are more important in the transition to sustainability than, you know, someone who catches the tube to London (…) every day. So I think, it never hurts to get to know more people who have a kind of different experience. (Empirical research p12-R1)

Evaluating the outcomes of the SuScit project, most researchers argue that a key outcome has been the direct dialogues and community engagement, which, by the nature of everyday life, transcend academic disciplines and categories, and which the researcher reflectively could take on board in various ways. While most professionals were conscious that the prerequisite for doing so is genuine engagement from both sides, a main concern is the question of how research communities can adequately take part in these modes of engagement without merely using people as a source of knowledge. One of the local residents reflects on this evaluating the project:

> I told you, I shared everything where I feel, I just told you, how it was, and it has to come back. I can say it was good experience with people around us, and I can see what we did, I can see some action around here. And now I've explained, and this is good. And I hope … let's do more things about this, no I don't mean for the state, but for the people who (…) are going down (…) like this. (Empirical research p12-LP8)

As part of the Researcher and Practitioner Workshop researchers themselves identified a whole range of fundamental institutional challenges to linking research and community engagement, including:

1. The logic of initiating research at the university and thereafter building in dissemination makes it hard to involve the local community at an early stage in building partnerships and developing shared goals for the research projects;
2. Ensuring that research also delivers practical benefits for community participants is rarely seen as a direct outcome of the project, and there is often a mismatch in modes of funding available for research and what is required for facilitating effective community involvement;
3. Academia does not have attractive researcher incentives or provide recognition for non-academic research outputs such as outputs other than those published in peer-reviewed journals. (Empirical research p07)

Such challenges clearly indicate that furthering public engagement is not merely a question of developing new participative methods and approaches. It is in particular a challenge of changing the structural and institutional conditions in and outside academia which frame the epistemological conditions for knowledge production.

Hence, building on the various perspectives brought up in this chapter, there is no question that a social orientation emerged amongst the researchers in the project, in terms of improved reflexivity upon community issues and how these related to research. However, in terms of actions taken forward, it was also clear that researcher reflexivity tended to be more focused at research perspectives than the community action. Here the incentive structures of academia, identified by the researchers themselves, seem of particular importance because they seem to impede the possibilities for researchers establishing the sense of responsibility towards community issues. However, without a shared sense of responsibility it is difficult to build momentum for creative processes of social imagination to emerge, and collaboratively start thinking and acting differently upon the shared issues brought up. In this sense many of the ideas of the project can be seen as initial stages of social imagination: identifying problems, challenging perceptions, sharing perspectives, whereas in terms of social *imagination* – thinking of alternative ways of living in the future and how research and community action could jointly enable change in such direction – the development of new research ideas became, for good reasons, more focused at translating and relating these perspectives into current frameworks of doing research. My point here is not that aspects of social imagination was not taking place amongst the researchers: in fact many of the research proposals developed by researchers through the SuScit project can be seen as building on and translating processes of imagining alternative futures into academic language. My point is that the community part of the project in particular made a strong basis for orientations towards social imagination. In this perspective the overall bottom-up structure of the SuScit project, starting with the community film-making process followed by shared workshops for researchers and practitioners to listen, learn and respond by developing a community-led research agenda, seem highly useful in a social learning perspective. The question that arises, however, is how far imaginative ideas of such processes can reach into the world of research, where institutionalised structures and everyday practices of doing research provides the epistemological frameworks for working as a researcher.

Social Learning as a Social Process

What can be learned from the above social learning perspective on the Citizen Science for Sustainability project? As it turns out the most interesting aspect of this analysis is not to evaluate to which degree social learning took place. The focal point of the above analysis is that once applying a social learning perspective a number of tensions become visible. Of particular interest for the focus of this book, one of the common themes in these tensions is grounded in the researchers' own reflections on the nature of doing research, and whether that adequately responds to the community voices articulated through the project. This was particularly clear in the evaluation interviews with the researchers involved: through the community engagement the project established a different kind of arena, inviting for reflections on research itself. Thus, before rounding off this chapter, let me go a bit more into detail on this particular issue.

One of the fundamental tensions, which can be found in the researchers' reflections on the project, concerns the question of whether current modes of doing research are adequately tuned to tying up with the dynamics of community engagement. As already shown, I prefer to describe this as a tension, because no homogeneous perspective was to be found within the group of researchers. However, among those who actually found it relevant to link research and community engagement, doing so in particular addressed the question of how research can actually do so (Empirical research p12- R2;R6S;R7;R9;R12). This was a theme going through several of the evaluation interviews with the researchers: *you don't want to treat people like a laboratory* (Empirical research p12-R9 and R2;R6S;R7;R9;R12).

What became clear through the Shared Workshops was that in the social meetings between researchers and residents, the community was not only an object of study with relevance for academic research. It was a concrete community inhabited by real people, with needs, hopes and desires. Involvement in this social space called for commitment from both sides. Many examples can be highlighted to illustrate this aspect of the project: in the beginning of the project concrete concerns over the urgent lack of housing in the community were taken up and acted on by researchers, not just as a research issue but as a moral call for action for those who came to know of this problem. Later in the project a key concern among the researchers involved appeared to be the feeling of having an obligation to respond and feed back to the community in ways which could have real and tangible value to the lay-people involved. By the end of the project a key concern of several researchers was how research can link up with this kind of community engagement. In my perspective, the implication of perceiving the community engagement being a social process is that it is not adequate to describe the dynamics between researchers and lay-people in terms of the rather simple subject-object relation of the observing researcher and the researched community. But what is the alternative? In my perspective, the researchers' reflections on the project can be seen as articulations trying out what that relation *could be*. More than forceful statements, these articulations appear as search processes questioning the way research is being done. Whilst I do not argue that the project established a fully unfolded social learning process, I find that applying a social learning perspective shows that shared social arenas emerged through the project, enabling participants to address and reflect on the way research is being done, and whether those orientations of science appropriately meet the challenges of urban sustainability that were addressed. In other words, the project as an experiment became an opportunity for highlighting the tensions between, on the one hand, the dynamics of research and, on the other, the dynamics of community engagement. This is why the SuScit project holds an exemplary potential for researching this particular aspect of upstream public engagement. In the following chapter I shall put these specific experiences into a broader perspective.

Chapter 8
Action Research for Sustainability

In this book I have sought to address the overall research question of how community-based action research in the area of upstream public engagement can further new research orientations towards sustainability. Building on the previous theoretical, empirical and analytical perspectives, the purpose of this chapter is to discuss four distinct questions. The first is: What can be learned from examining the Citizen Science for Sustainability (SuScit) project building on theoretical approaches to sustainability in the context of democracy? The second: What are the contemporary conditions for furthering these perspectives in new modes of knowledge production? The third: What is offered by the framework of Critical Utopian Action Research in this respect? And finally: What do these considerations imply in terms of the idea of science furthering sustainability? To me, considering each of these questions is crucial for examining the overall research question posed by this book.

Learning from Theoretical Concepts of Sustainability

What theories enable us to do is, for a time, to see the world from a certain perspective. Whether that proves useful or not depends on the actual synergy, which might emerge in combination with the particular empirical field: does it expand our insight or place limitations on it? Answering that question takes both listening and critical triangulation of theoretical and empirical insights; simply forcing theory onto empirical experiences neither adds value nor validity. In the following I have sought, critically, to test conceptualisations of sustainability in the context of democracy (outlined in Chapter 5) against my empirical experiences from the SuScit project (as analysed in chapters 6 and 7). I guess the process can best be described as active listening in order to critically answer the question: In what way, if any, does it make sense to apply this theory to the actual empirical experiences? Obviously this neither serves as an absolute validation or falsification (neither is my purpose), but it explores whether insights might emerge from combining the two.

A first insight that emerges is that the articulations of local issues, which are *not* perceived as sustainable, turn out to be a core aspect of the SuScit project. The *community identification of un-sustainability* was a focal point of the process. This was the case for young people addressing the social and physical lack of spaces to be; the elderly in their ambiguous call for love and friendship; the lone women in addressing challenges of modern life through the community garden work (and several similar examples). Taking a closer look at these examples it appears that most of these issues are social challenges more than environmental. Or, in Shiva's terminology, challenges to the sustenance economy seemed to appear as more pressing in the context of urban everyday life than challenges to nature's economy. This does not mean that environmental concerns were not to be found in the everyday life context. But it reflects the fact that in modernity the human–nature relation has increasingly become managed through socially differentiated expert systems, rather than being more directly attached to urban everyday life. On the contrary, essential parts of the sustenance economy (raising kids, being part of a neighbourhood and so on) are much more visible in everyday life, because they *are* everyday life. In modern urban life, citizens have become much more indirectly connected to and dependent on nature's economy, compared to the sustenance economy. In this sense it was perhaps not surprising that the experiences from the project in particular made evident that *social* sustainability is threatened or under erosion. However, it also became clear through the community work that in everyday life nature relations are embedded in the social. Perhaps the clearest example is the community gardening, where elements of nature were part of coping with social challenges of deprived urban living. But similarly the issue of lack of spaces and calls for community cohesion both had social and physical dimensions. Most community participants perceived 'nature' as an asset of everyday life, for example by mentioning green space, clean air, the light of the morning sun and so on as quality of life aspects of their urban living (Empirical research p04-OP). Clearly the fact that 'nature' is part of the social does not necessarily imply any sustainable relation between the two. But it is an important feature of the citizen dialogue that in everyday life perspective, notions of nature and the social are highly interrelated. In fact it seems that one of the true potentials of the citizen-expert dialogues was to explore this interrelatedness. In Chapter 6, I highlighted a dialogue of a man who wished for a huge house as an example of how material aspirations and social needs can be interlinked, and that unpacking them might help explore alternative solutions for future sustainability. My point is not that all material aspirations are merely social needs; that certainly is not the case in a deprived urban area facing substantial economic challenges in relation to daily living. My point is to seek for concrete ways to integrate environmental and social sustainability.

Let me elaborate further on this point by reconsidering the example of the community garden. In a societal perspective, the garden can be seen as representing a counter-initiative taken by women empowering themselves to be less dependent on societal structures associated with social deprivation. But at the same time, using Shiva's terminology, the community garden represents a potential of re-embedding a very direct nature-relation into a local citizens' sustenance economy. In this perspective, the exemplarity of the community garden is that it inherently holds the potential for integrating environmental and social sustainability. We certainly know that there are conflicts between the interests of nature and society – if not there would be no need to discuss the issue of sustainability. However, the challenge is not merely to identify this historically constituted conflict, but to find ways to overcome it. That is why examples which might provide solutions to environmental *and* social concerns are of particular interest. Recalling the researchers' responses to this, what seemed to happen in the process was that researchers started 'translating' community issues into the language and logics of their various research fields. This is a crucial point: *the orientation of research starts becoming responsive to local community issues*. My analysis exemplifies that this was done in several different ways: by suggesting developing applied technical solutions in direct collaboration with the community and its own identification of local needs; by suggesting social research for foster local community empowerment; and by suggesting critical analyses of the societal implications of the ideas brought up by the community. Thus researchers responded to the community dialogue at technical, social and at societal levels. Obviously more traditional research approaches were also represented within the project. The interesting aspect of these examples however is that they showcased how researchers were using their analytic capabilities and professional experience to 'lift' local issues into various socially differentiated spheres where they could be addressed and potentially responded to. Whilst many of the community ideas can be seen as local counter-initiatives to societal structures of social deprivation, researchers were in a somewhat privileged position to highlight that the issues raised could be addressed at various different levels in relation to the way societal systems operate. In a modern out-differentiated society this is obviously important to enable change.[1]

Seen in a more general perspective, perhaps the most novel aspect of the SuScit project emerged from – more or less successfully – serving as an alternative epistemological space, whereby marginalised experiences of urban deprivation were put on the agenda, thus revealing challenges of unsustainability, and providing an opportunity to start reflecting on how these could be met by the integration of environmental and societal insights. This reorientation of research was far from without problems. Whilst I shall come back to some of the challenges in the following chapter, for now I shall take a more detailed look at what I see as the potential of this process.

1 It should be noted that such issues are complex: for instance gardening projects, aiming to empower deprived neighbourhoods, might be attracting more affluent residents, in effect *consolidate* processes of gentrification.

Probably the clearest expression of this tentatively emerging reorientation of research was caught in the theme *you don't want to treat people like a laboratory*. This was a challenge raised by several of the researchers in the evaluation interviews (Empirical research p12-R2;5;6S;7;9;12). The task is not to enter the local area, grab the resources for academic knowledge production and return as if nothing had ever happened. The purpose is also to support sustainability, not least *in* the community. In a theoretical perspective, this distinction has some very deep implications with parallels to the framework provided by Shiva: her interpretation would be that the purpose is not the enclosure of local knowledge (for the gain of a privileged academic community) but to increase sustainability, through the democratic potential of knowledge that its value increases by being shared. In this perspective, *sustainable knowledge* consists of insights which support people in actively maintaining the balance between nature and society. I mention this particular theoretical perspective because I think it might help us understand the somewhat tentative questions and potentials emerging from the experiences of the SuScit project. In the complexity of the modern out-differentiated society, sustainability has little chance of emerging and unfolding from everyday life without changing the societal structures, which interdependently constitute the un-sustainability of the basic human actions of everyday life. But everyday life must be able to deal with these changes, and a prerequisite for doing so is that these are inherently environmentally and socially sustainable. In my view this implies a new horizon and categorical imperative challenging science: *to sustain the sustain-ability of life*. In this conceptualisation I strongly draw on what I think essentially can be learned from Shiva's thinking, namely that the very aim of research must be to support people and communities in sustaining themselves while preserving nature's own ability to regenerate. In my perspective this is where sustainability becomes sustain-ability, and where new orientations of science might have a role to play.

Understanding Dynamics of Knowledge Production

How do these ideas about science and sustainability in the context of democracy correspond with the actual dynamics of knowledge production? In the first chapter of this book I introduced the notion of mode-2 science as one of several contemporary ways to conceptualise what is seen as an ongoing historical transformation towards more contextualised modes of knowledge production (Gibbons et al. 1994; Nowotny, Gibbons and Scott 2001). Now it is time to examine this conceptualisation more critically by drawing on mode-2 literature, as well as a number of contrasting perspectives, in order to discuss whether the ideas put forward in this book could in fact be part of such new modes of knowledge creation.

Despite the common starting point – the idea that the nature of science is and continuously will be changing throughout history – it should be clear that the perspectives put forward in this book in many ways differ from mainstream discussions of new modes of knowledge production. The first difference concerns what is actually meant by *contextualisation of knowledge*. In the framework of mode-2 theory contextualisation implies that

> … society is able to "speak back" to science, and that this reverse communication is transforming science. Contextualization is invading the private world of science, penetrating to its epistemological roots as well as its everyday practices, because it influences the conditions under which "objectivity" arises and how its reliability is assessed. (Nowotny, Gibbons and Scott 2001: 54)

However, as influential the mode-2 theory has been in arguing for the emergence of new and more contextualised modes of research, equally tentative it has been in providing more exact sociological concepts of these processes of contextualisation. Whilst the literature on the philosophy of science provides profound insight and understanding into the way in which mode-1 science continuously develops criteria for good science through paradigmatic battles among academic communities, the corresponding criteria for the validity of mode-2 science is the rather vague concept of 'socially robust knowledge'. On the one hand it is argued: 'Reliable knowledge can become socially robust knowledge only if society perceives the process of knowledge production to be transparent and participative' (Gibbons et al. 1994: 248–9). On the other hand it also emphasised that 'Mode 2 involves the close interaction of many actors throughout the process of knowledge production and this means that knowledge production is becoming more socially accountable' (Gibbons et al: 1994: VII). However, in reality these ideas of transparency and participation are not as straightforward as they might seem at first sight. The implication of this way to conceptualise contextualisation is that the validity of mode-2 science becomes highly dependent on its impact among various societal actors. However, without more explicitly defining this interdependence, as well as profound criteria for the relevance and validity hereof, mode-2 theory easily implies that any external interference can be seen as 'contextualisation' and thus as contributing to the production of 'socially robust knowledge'. As argued by critics, the concept of mode-2 science might in reality turn out to provide a black-box-theory distinguishing, rather than making transparent, the way in which science actually becomes increasingly dependent on and influenced by strong external stakeholders (Kamara 2009). Thus, the vague definition of contextualisation potentially implies that the mode-2 ideal of socially robust knowledge turns into what could rather be termed *commercially* robust knowledge, because the 'context' in reality turns out to be dominated and defined by strong external stakeholders with preset interests. Hence I would argue, for mode-2 theory to offer more than arguments for increased commercialisation of science, more exact concepts of, criteria for and methodological approaches to 'contextualisation' are needed.

It is in this perspective that this book offers an approach for developing more exact sociological concepts of contextualisation, why it is needed and how it might take place. My particular aim has been to understand the challenge of sustainability, not only in terms of increasing environmental challenges, but also by approaching sustainability as a societal and cultural issue. From my theoretical analyses it seems evident that the present state of unsustainability is not least related to the dynamics of modernisation. If so, an essential part of the challenge of sustainability is that systemic rationalities need to be transcended. In a critical theoretical perspective this implies building on the human lifeworld in order to enable rationalities, which are not systemically predetermined. Hence, a key question of this book is how we can enable new modes of science to do so. The emerging insight is that the contextualisation of science must open up to the human lifeworld, a concept which interestingly is not part of the mode-2 theory. Throughout the book I have sought to explore and highlight a number of ways this can actually be done. Rather than using the somewhat vague mode-2 concept of 'contextualisation', I find building on Shiva's thoughts that what is needed is a *reorientation* by which the aim of sustainability science is to support modern ways of living, capable of sustaining itself without eroding the social and ecological relationships by which it is constituted. That would be a sustainable reorientation re-embedding science in its societal and ecological context.

What are the possibilities of such new research orientations taking place? In my analysis I have examined this question at a rather empirical level, showing some true potentials as well as clear barriers in this respect. In the following I shall try, at a more general level, to address these barriers and what implications they might have. My starting point, in relation to mode-2 theory, is to re-examine what is actually meant by the concept of knowledge production. To enable a critical triangulation of this concept I have found it useful to contrast mode-2 theory with Boltanski and Chiapello's discussion of *The New Spirit of Capitalism* (2005). In this perspective the term knowledge production obviously implies the production of knowledge, but it also has more wide-ranging societal and historical connotations. The term does not merely cover the epistemological process of knowledge *creation*, but knowledge *production*, echoing the ongoing historical development from industrial society to knowledge society, in which knowledge is increasingly becoming part of societal *commodity production*. When science is seen as societal commodity production, the value of research is not least estimated by its input into the market economy. The implications of this development are evident inside academia in terms of new economic steering mechanisms and ways of organising, incentivising and structuring research (Greenwood 2009) But the idea of turning knowledge creation into market value also has deep implications in the 'societal context'. Whilst scientific knowledge production certainly does create new knowledge and impressive insights, its societal role is much more contested. Kristensen (2008) reflects on the nature of knowledge and its role in society:[2]

> Knowledge is not naturally a *limited resource,* which value is the fruit of its natural scarcity. If anything knowledge is by nature a common good, the production of which is the result of implicitly collective and corporate processes – and an asset, which increases by being shared with others. Thus, the economic value of knowledge can essentially only derive from an *artificially established scarcity* in forms of institutional, political and/or power-based constrictions to the *access to knowledge and use of knowledge* as a fundamentally common good. (Kristensen 2008: 104 *my translation*)

The prerequisite for research being a driver of the *knowledge economy* is that restrictions and criteria for what is counted for economically valuable knowledge are established. This logic per se implies that knowledge, to which everybody has access, can hardly be counted for as economically valuable. The implication hereof is that everyday life knowledge and lived experience has only very limited economic value as long as it can hardly be capitalised (or in Shiva's terms: enclosed). As Kristensen notes, the 'paradoxical then is that knowledge first and foremost generates value when being disseminated and socialised, while dissemination inversely reduces the possibility to dedicate, privatise and capitalise it' (Kristensen 2008: 104 *my translation*). In a sustainability perspective, this effectively means that cultural processes taking place among citizens in their everyday lives, providing an essential foundation for social sustainability, are not counted for any economic value if not enclosed into commodities in the knowledge economy. The enclosure of knowledge implies that increasing scientific insight does not necessarily contribute to citizens' possibilities to further societal or ecological sustainability. Among several examples of this, Shiva has highlighted how commercially patented, genetically modified annual crops, developed by knowledge-intensive companies in other parts of the world, are replacing and eroding cultural knowledge for biologically diverse community farming, resulting in ecological and social instability (Shiva 2005). The implication of this dynamic is that science both holds the potential to

---

2 Through a comparative reading of Boltanski and Chiapello (2005) and Boutang's thesis of cognitive capitalism (2007).

strengthen and to erode sustainability. As phrased in Shiva's terms, increasing efficiencies and great growth in GNP (in one part of the world) can happen on the basis of enormous losses of nature's economy and the sustenance economy (in other parts of the world). Hence, it is clear that the production of knowledge and new insight does not per se further sustainability; it inherently holds this *potential*, as well as the opposite. A key challenge to new modes of research, which are increasingly becoming drivers of the societal market economy, would be to find alternatives to the ever-increasing dynamics of enclosure. To put it simply, the enclosure of knowledge does not further sustainability because the basic aim is not through knowledge creation to further the ability of nature and society to sustain itself, but to expand the academic system.

Insights Offered by Critical Utopian Action Research

Taking the above considerations into account, how can community-based action research help science further sustainable change? Examining the experiences from the SuScit project reveals that the project both offered important potentials as well as barriers. Many of these are not unexplored land, but have received in-depth discussion in the field of action research. In the following I shall highlight a number of ways in which I think the framework of Critical Utopian Action Research might strengthen some of the experiences from the SuScit project towards furthering sustainability.

Action research today represents manifold well-established research fields, each providing specific approaches for various purposes (Reason and Bradbury 2008; Greenwood and Levin 2007; Nielsen and Svensson 2006). In the following I shall particularly draw on the research tradition of Critical Utopian Action Research developed by Kurt Aagaard Nielsen, Birger Steen Nielsen and Peter Olsén, especially in relation to Future Creating Workshops and Research Workshops (1999; 2006a; 2006b; 2007). The particular strength of this approach, as opposed to pragmatic and systemic action research offering valuable insights into organisational and systemic change (Ramage and Shipp 2009; Blackmore 2010), is that it critically addresses the challenge of moving beyond the logic and rationales provided by pre-established systems. Theoretically, this particular orientation links with a specific challenge of furthering sustainability, namely to overcome sustainability as a purely systemic concept. Recalling the notion articulated by Elling:

> The criteria for a new agenda for sustainability must be that the systemic rationalities as well as those rationalities from the lifeworld are taken together and made the basis for actions (…). We must open up the possibility for other

> rationalities than systemic rationalities, and we must open up the possibility for other orientations of actions than goal orientation. (Elling 2010: 39)

This challenge, however, is not merely philosophical. It is a practical real-world challenge, and this is where Critical Utopian Action Research might offer valuable insights. Particularly I find it useful for examining and obtaining a deeper understanding of the social arenas, which were created through of the SuScit project. Elaborating particular on this issue is the purpose in the following.

The Creation and Doubling of Free Space

In my analyses I have already noticed that a particular feature of the SuScit project was to enable a social space of community work as the outset for a bottom-up engagement process between residents, practitioners and researchers. Rather than grounding the process in pre-existing academic discourses, the aim was to allow community members themselves to explore and share experiences of what it was like to live in their local area through the means of community film-making. My analyses of this process evidently show, on the one hand, how this approach both produced perspectives which did not fit very well into established academic discourses of sustainability, but, on the other hand, how this implied the possibility of starting to think outside the box and explore issues which are highly related to urban challenges of unsustainability, although not necessarily high priority in contemporary academic deliberations. Due to the reflections of the researchers taking part in the project, a particular intellectual value of the community perspective was not just empirical experiences fitting well into theoretical categories, but in fact everyday life perspectives *challenging* scientific thinking and underlying assumptions. Thus, an important finding is that perhaps the biggest potential of the community approach is that it truly represents and articulates something different than academia often does: the experience of lived life.

To understand this particular feature of the project, the action research concept of *free space* appears particularly valuable. In Nielsen and Nielsen's conceptualisation, the idea of free space is to foster social arenas in everyday life, where authoritarian social structures of reality-power, constricting people from thinking and speaking freely, are delimitated. To claim that free spaces are power-free is impossible (for example taking into account Foucault's insights on the nature of power). But it is a social arena where dominant external power structures are delimited in order to enable peoples' everyday life perspectives to be articulated more freely (acknowledging that these are still partly framed by those structures). The social and epistemological quality of establishing free spaces in everyday life action research is that what is normally being suppressed or marginalised by the contemporary societal power structures, in which everyday life is embedded, can potentially be articulated and shared. Such issues are often socially sensitive and highly ambiguous. Free space is not something which can just be socially installed or forced through. Rather, on the contrary, whilst one-dimensional goal-orientations will always limit the emergence of free space, the exclusion of formal power relations and facilitation to create a trustful social space are among the practical ways to its establishment. At a methodological level this approach has particularly led to the development of Future Creating Workshops as one specific action research method (Jungk and Müllerts 1987; Nielsen, Nielsen and Olsén 1999; Drewes 2006; Nielsen and Nielsen 2006a; 2006b). From numerous action research projects it has become evident that when free spaces are successfully established, they *call* for marginalised and suppressed voices to be articulated and shared. In this sense it is not surprising that if the community work of the project more or less successfully established free spaces, the articulation of what I have interpreted as unsustainable features of modern urban living was a result. In an action research perspective this is in fact a successful outcome of a process establishing free space.

The concept of free space offers a specific framework for understanding why community-based public engagement can actually add important insights to further sustainability: because it makes room for articulating what is perceived as unsustainable in people's everyday lives. However, the notion of free space became useful in a *dual* sense: both as a free space in citizens' everyday life and as a *free space in an academic context*. In the SuScit project the free space was, although not free from, at least less dependent on being pre-determined by academic rationalities, discourses and power structures. In this sense it was not merely a free space for the residents, but to some extent also for the researchers involved because a different arena for deliberating on sustainability emerged. The approach that researchers and practitioners were invited, not in the traditional role of giving expert advice, but to listen and learn from the perspectives emerging from the community work, made it possible for them, to some degree, to take part in free spaces to which they normally would not have access. This is because these were not least constituted by the delimitation of societal power structures particularised in professional roles and expert functions. It was particularly noticed by several of the researchers and practitioners during the workshops that establishing this level of community dialogue would normally be beyond the scope of professional practices. Some of the professional participants also actively reflected on this, noting that it was interesting taking part in a social process without being in the role of formally representing institutional interests (Empirical Research p12-R2).

Based on my analysis, I suggest that this particular feature could be conceptualised as the *creation and doubling of free space*. Two particular features of the process appear to me to have actively contributed to this doubling of free space. The first was the way in which academics and practitioners were invited, introduced and instructed to take part in the project. Based on the idea of developing a bottom-up community-led agenda for urban sustainability research, it was seen as essential in the project facilitation that the role of academics in the shared workshops with residents was not to be experts. Rather this was a chance for them dialogically to listen and learn, whilst residents could express their views on what local life was like. This instruction was given to all participants by the facilitators. The intention of so doing was to counteract pre-existing academic discourses framing the entire dialogue, only leaving space for community perspectives as long as they fitted into academic categories. In other words, the academic primacy on defining and understanding sustainability was intentionally 'disturbed' through the process-design and facilitation. In the actual process this did far from implying completely to exclude or overrule the complex social dynamics associated with ordinary residents from a deprived urban area meeting high-ranked internationally acknowledged senior academics. But the fact that the project organisers through the facilitation actively sought to redefine the expert lay-people relation as a premise for taking part in the project at least called for a relatively open and less predetermined approach in the meeting between academics and residents.

Another feature, which turned out to be vital to the process, was the use of arts in terms of community film-making. The idea of residents themselves developing community films over a longer period of time, before presenting to the academics as the very foundation of initiating the shared dialogue, effectively helped setting the agenda from the community perspective. Showing the films on large screens at the Shared Workshops powerfully brought to life challenges, aspirations, ambiguities and paradoxes experienced in everyday urban life. The films in many ways did not fit into strict academic categories of sustainability. But as films they were perceived as coherent pieces of art with their own legitimacy.[3] Through the films, stories could be told, which were not simple, clear-cut messages, but expressing and showing the ambiguity of everyday life. The essential value of them was that the community perspectives were brought into the project in a much more complex and multifarious way than would have been the case if residents had been asked for example to write

3 An important aspect of this use of films is to consider their epistemological legitimacy. For the researchers and practitioners involved, the films became a representation of particular social urban realities, but thereby also reproductions hereof, one the one hand, communicating everyday life issues in a very direct way, but, on the other hand, also adding an extra 'filter' between the film-makers and the audience. While the films were highly useful for initiating a dialogue grounded in everyday life perspectives, they should not replace the direct dialogue to gain insight on the everyday life issues at stake.

down a list of key issues of modern urban living. The films were the residents' films, produced by them, representing their issues in the way they had wanted to do it, and therefore effectively brought the residents on stage, because they were the ones having the ownership to this very starting point of the collaboration with the practitioners and researchers.[4] Hence, I find that the essential value of the films was that they were part of suspending traditional academic agendas as the basis for the shared deliberation. To do so, community ownership and the use of arts turned out to be crucial.

Compared to contemporary public engagement methods, it is worth noting and reminding organisers of such initiatives of the importance of considering two essential questions: Who owns the discourses initiating and framing the deliberation? And what languages are perceived as valid in the dialogues? The particular value of art in this context is that it is one of few languages suitable for accommodating paradoxes, antagonisms and ambiguities, which are not always suitable in academic argumentation simply because it implies the risk for rationalistic and scientific arguments to implode. This is a particular feature of art, which has been extensively elaborated in critical theory (Adorno 1998) as well as in practical action research (Tofteng and Husted 2011; Brydon-Miller et al. 2011), evidently showing that the use of art should not be reduced to endless inventions of attractive and creative toolkits, but be methodologically acknowledged for its epistemological value: to open up and transcend established ways of thinking. The point here is not to erode scientific thinking. The point is that in order to open up academic discourses towards understanding and taking notice of the lived experience of everyday life, it is necessary to include ways of communicating, suitable for expressing these perspectives and understandings of life. Considering the experiences from the SuScit project, it seems that to enable a free space in a field of academic discourses, the orientation towards everyday life and use of art, as alternative means of communication, was essential. This being said, however, it is important to notice that these must be understood as temporary free spaces in a broader institutionally framed academic landscape, and the consequence hereof was a conflict between the two.

To understand this more in depth, it is useful to consider a few more action research concepts closely related to the notion of free space. As already noted, an important feature of the SuScit project was that something emerged through the social process which could not be reduced merely to a product of the residents, practitioners or researchers separately. The clearest indication

4 Methodologically and ethically it should be noted that using community film-making as the basis of sharing vulnerable aspects of everyday life represent a particular challenge, for which a trustful social space and clear and ongoing participant agreements on ownership, purposes and potential audiences for different parts of the work seem to be particular important.

hereof, perhaps, is captured in the notion that *you don't want to treat people like a laboratory*. Why not, one might ask? There is a lot of profound social research making coherent observations building on subject-object relations with clear epistemological similarities to classical research lab set-ups where people *are* being treated like laboratories. In my view the notion – that you don't want to treat people like a laboratory – is a social and ethical question confronting the role of science and its research methodologies. Considering why this was the case I think it is useful to recall that the SuScit project was not merely an invitation for social observation, it was an invitation to collaboration and community engagement. To understand this difference the concept *the common third* might be useful, highlighting that in a research process involving real people there is always more involved than what is at stake for the researcher and the participants respectively. The common third is 'the product of co-operation between the subject (the researcher) and the other (the participants)' (Tofteng and Husted 2006). Whilst this can be said (although not always acknowledged) to be the case for any social science research, in action research the common third obtains a particular role. Here the common third is not just an arbitrary 'social construction' (as many dialogues appear to be if only analysing them at a discursive level); it is an already existing common among the participants, called into being. Such commons can be very concrete aims, tasks, processes or products co-evolving between researchers and participants, thus establishing a new interdependent mode of collaboration, not being completely controlled by either of them.

In my view, if a common third was established in the SuScit project, it was in terms of a shared commitment to a particular type of process, an experiment, which *could* turn out to approach local concerns differently than often seems to be case. The notion that you don't want to treat people like a laboratory is interesting in this light, because it builds on what could be a common third: to further local sustainability. Such orientation can partly be seen as a common third in the SuScit project. But it largely remained a potential feature of the project, *calling* for new orientations in shared collaboration, but not fully evolving into practical steps of realisation. Paradoxically, the constituting framework of developing a community-based research agenda not least implied a somewhat streamlined process towards reaching this aim rather than providing participants themselves the time and space to explore and identify their own common third, which could have been called into being. To me, and as acknowledge by many other thinkers (Cooke and Kothari 2001), this problem is rather common in public engagement activities, often being predetermined by preset aims rather than calling for mutual collaboration. Hence, the concept of the common third at least highlights one particular feature of the notion of free space, calling for new types of commitment and engagement, which might be established in future projects.

To understand the role and relevance of academics in this context I have already empirically analysed the researchers' involvement in the SuScit project in terms of the concept of *free distance* introduced by Nielsen and Nielsen. Building on this analysis it is worth noting that free distance becomes particularly important in relation to the discussion of free space. Free distance implies the possibility at the same time to engage in and critically examine the collaboration process. Without engagement, researchers establish a social distance to the other participants thus constituting an epistemological barrier for understanding the issues at stake as experienced in everyday life. But without distance, researchers risk becoming 'victims' of whatever community perspective the process might foster (the same is the case for all other participants). Hence, in Nielsen and Nielsen's conceptualisation, participants' free distance – moving between distance and engagement – is a basic prerequisite for enabling free space.

In my analysis of the researchers' reflections on the SuScit project, I have already highlighted that reflections on the way research is currently being done, and how it corresponds with engaging with local communities, turned out to be a key outcome of the process. To me the notion of free space and free distance offers particular value in conceptualising this feature of the process. In my interpretation, what effectively happened was that through the doubling of free space researchers were prompted for free distance, not only to the community perspectives, but also to academia itself. The analysis in Chapter 7 in particular shows that many of the researchers' reflections on the process were indeed self-reflexive, considering how research is actually being done and how that corresponds with furthering sustainability in urban communities. Taking into account that many of the researchers had former experience in public engagement, my point is not that the SuScit project alone produced these reflections, but rather that the project effectively provided a social space for articulation and reflection on this issue. Hence, my interpretation is that the process somehow called for reflexivity.

To understand this particular aspect, a brief excursus from the action research perspective is useful, recalling the discussion of Chapter 5 on Elling's distinction between socially mediated reflexivity versus systematically mediated reflexivity:

> Reflexivity on the basis of the lifeworld may then be described as a process that aims to reach understanding of something – its origin and consequence – in the outer world, the social world or the inner world, in which this understanding is realized through a combination of communicative action, normative foundation and cultural legacies.

> [R]eflexivity with a systemic form of organization must be characterized as a borderline case that concerns only the outer world, and in which the teleological and strategic forms of action, with their corresponding forms of rationality are decisive (Elling 2008: 197–8).

Based on these definitions, it is possible to conceptualise more precisely the notion of reflexivity. From my analysis of the project, my conclusion is that the process in fact called for socially mediated reflexivity, thus confronting the systemically mediated reflexivity of academia. The researchers involved, both being part of the academic system and taking part in the community engagement, can be seen as mediating between these two forms of reflexivity. This was particularly visible in researchers' considerations over the value of the dialogue for academia and the local community respectively, as further exemplified in Chapter 7. My point is not that expert-reflexivity is normally purely systemic. My point is to highlight the conflict between the rationality, aims and requirements of the academic system, and the idea of engaging with local communities aiming to further local sustainability, as a common underlying theme in the researchers' reflections on the process. In the engagement process systemic rationality was confronted with lifeworld perspectives and lived experiences of everyday urban living. In my perspective this process holds a potential for developing forms of expert-reflectivity better corresponding with what is perceived by urban residents as important challenges to sustainability. If, as suggested in the introduction, a key challenge for moving towards sustainability is to enable social change by better understanding sustainability issues from a lifeworld perspective, this observation is potentially of great value.

It is in continuation of this consideration that I find the action research term *social imagination* particularly interesting, because it helps conceptualise the added value of the process. In Nielsen and Nielsens' framework, social imagination is an inherent potential feature of free space, enabled by the creativity emerging from it. Whilst I have already theoretically introduced this term and used it in the analysis of the SuScit project, at this point I shall merely highlight that social imagination, free distance, the common third and free space are highly interrelated concepts. Together I find that they might help conceptualise ways to enable shared imagination that is embedded in, and reflects particular socio-historical contexts (so as the challenges of modern urban living) without being determined by this scope. It is in this particular context that the notion of social imagination has a more radical take on what is meant by social learning. If social learning merely means building shared understanding across participants, it seems rather unlikely that the learning process will be able to transcend underlying structures in the relations between citizens and scientists. Social *imagination*, addressing the basic democratic question *how do we want to live*, potentially adds direction to the learning process. Since researchers do not per se have a privileged position to answer this fundamental question (although it might implicitly be part of research ideas), the learning process is potentially becoming oriented towards a broader democratic deliberation on the role of science in society as well as ideas for how science might respond to needs in the local community in relation to broader societal challenges. Thus social imagination holds a critical potential for initiating learning processes that do not simply reproduce academic thinking, but situate science in particular societal contexts.

I am building on these insights when I find that the concept of creation and doubling of free space holds an epistemological potential for knowledge creation. An important task exploring this potential could be to initiate new experiments gaining experiences in different contexts. In doing so it would be essential to acknowledge the doubling of free space as an intrinsic aim of the process from which new insights might emerge. Such development would be interesting in a number of different ways. One would be to experiment on doubling free space as part of upstream engagement initiatives. The consequence thereof would certainly be in contrast to older public engagement methods, orienting upstream public engagement towards new kinds of collaborative knowledge creation. But the concept equally holds potentials for furthering the tradition of Critical Utopian Action Research itself. The perspective in doing so would be to experiment on methods, which do not merely establish a free space, for citizens to articulate everyday life perspectives, but equally enabling a double free space, in which organisers and action researchers themselves took part as an epistemological outset for gaining new insight. These suggestions obviously call for profound methodological consideration and further development. Through the examples and perspectives of this book I have sought to highlight at least a number of insights which can be taken into account as inspiration for so doing. However, furthering these ideas *is* a question of future experimentation. It is in this horizon of social learning that community-based action research, in my view, holds a true potential to further new research orientations towards sustainability.

From Knowledge Production towards Knowledge Democracy

Based on these considerations, how can we understand how knowledge creation can further sustainability in the sense it has been conceptualised in this book? In my view this at least implies: 1) considering what is actually meant by the notion of sustainability; 2) thinking through what this means in terms of academic methodological practices; and 3) examining what this means for the relation between science and society. While these questions clearly call for practical experimentation, I shall briefly try to address my perspective on these as an outline for further development and experimentation.

1) Science in the Role of Sustaining Sustain-ability

If sustainability is in fact an immanent and emergent *ability* of ecological and social life to continuously renew itself without eroding its own basis for existence, sustainability in principle cannot be *invented* but only *supported* by science. The implication is that any science aiming to further sustainability needs continuously and critically to question in which ways it might potentially erode or sustain the sustain-ability of social and ecological life. Obviously this question is difficult, not least taking into account the complex socially differentiated structures of modern societies, to which further levels of complexity are added by the worldwide and cultural interrelations created through processes of globalisation. These levels of complexity make it incredibly difficult to obtain any clear answer to the above question; in fact ambiguities and antagonisms will probably always be the case. However, this should not prevent us from considering this question, which *does* imply a different role of science, than often taken for granted. Historically speaking, challenges of modernity are clearly different from those in which the seventeenth-century scientific revolution was embedded. The challenge is no longer merely to produce rational knowledge replacing ignorance, cultural and religious dogmas. Rather it seems that there is an increasing need to overcome the challenges which have emerged in the aftermath of the increasing knowledge creation of recent centuries itself. Thus aiming for sustainability is not least a question about the role of knowledge in this world, and it is in this sense that the above orientation critically implies a new categorical imperative for science furthering sustainability.

2) Expert Reflexivity and the Double Orientation of Research

What does this actually mean in terms of the way researchers operate and science is being done? Does this approach offer an *add-on*, an extra feature to be put on the top of science? Or is it rather an *alternative*, a truly new paradigm for research? Possibly the most convenient interpretation might seem to perceive the above as an add-on to already existing modes of science. The problem of so doing is that it is hardly going to challenge any of those underlying structures questioned above: on the other hand, perceiving these thoughts as completely incommensurable to contemporary research does not hold much potential for change. Examining history of science after all, the idea of a sudden transformation of one mode of science into a completely new one does not seem very plausible. Similarly, one could argue, dichotomising between mode-1 and mode-2 science seems illusionary, as mode-2 science can hardly exist as science without grounding it in mode-1 science. Still, I would argue, recalling the empirical experiences from the SuScit project, these questions call for more than minor incremental changes in research cultures. The notion that *you don't want to treat people like a laboratory* is not just a question of science communication; it is an ethical and an epistemological question about the research and innovation process itself, because it concerns the type of knowledge and insight which is being created through various ways of doing research. Hence, from my perspective, the above questions need to be addressed as part of the research process itself. How can this be done?

To do so I suggest that science must build on a *double-orientation*: on the one hand, science must develop from being fixated on a particular subject. But on the other hand, it must also be able to openly consider its own basic presumptions and orientations. My point is this: whilst the first is the privilege of being a researcher, the latter must take place through democratic spheres. The principle of scientific double-orientation implies both a democratic openness and the scientific focus on a particular subject at stake. Epistemologically, Nielsen and Nielsen's hermeneutical principle of free distance provides a circular model for doing so: engaging with questions rising from societal contexts; reflecting on these through academic research; feeding back to societal contexts (critically and constructively); being open to what new questions emerge; and so forth. Such process can inherently include mode-1 as well as mode-2 approaches to research. The difference is that the arena, on which these questions are addressed, explored and discussed are opened from primarily having their locus within scientific communities, towards much more actively be part of societal democratic deliberations. In my view this would be to further new orientations of expert-reflexivity more suitable for meeting the challenges of sustainability.

3) From Knowledge Production towards Knowledge Democracy

The above considerations might question understandings of the relation between science and society. The challenges of modernity cannot be met by science alone. While researchers are obviously highly qualified in considering intra-scientific questions, science itself is not fully sufficient to provide adequate answers concerning the role of science in its societal context. Such questions *are* societal questions, and hence in democracies, *democratic* questions. The increasing role of scientific knowledge in modern societies only deepens the need not only for scientific consideration, but also for democratic deliberation on the societal role of science. The contemporary logics of knowledge institutions being drivers of the knowledge economy make this challenge even more complex. Thus ever-increasing the efficiency of knowledge institutions, in terms of producing new competitive outcomes for global markets of innovation, effectively becomes the way for knowledge institutions to obtain societal legitimacy and ensure economic existence. Obviously this raises the pressure for transforming knowledge into market value through the process of enclosure. This very dynamic deepens the challenge of sustainability, since the systemic aim of knowledge creation is not ecological, not societal, but the knowledge institutions themselves. Hence the dynamic of knowledge production becomes increasingly systemic.

This very challenge, however, is not based on the nature of knowledge and knowledge creation. It is based on the way in which knowledge production is being organised as part of the contemporary knowledge economy. There is no doubt that these dynamics have some very clear parallels to those already seen in industrial production. This is the case not least in terms of increased goal-orientation to improve efficiency quantified on the basis of a preset number of parameters, while externalities are ignored because they fall outside the systemic scope of consideration. If the shift from industrial society towards knowledge society implies that knowledge institutions take over these highly unsustainable dynamics of industrial production, an increasingly important challenge to sustainability researchers is to overcome the inherent consequences of those structures they are part of qua their involvement in the knowledge production system. In a critical theoretical perspective it is highly questionable whether the academic system of knowledge production itself should have the capability to overcome this challenge. Rather it seems to be an obligation to be taken seriously by any researcher concerned with sustainability, which might include, in ambiguous ways, challenging the self-referential goal-orientation of the academic systems that any academic per se must be part of in order to be in academia. This challenge has some very clear historical connotations to industrial worker-conflicts. Sustainability researchers need to demand that the purpose of doing research also lies outside research itself. Or, at least, that the quest for knowledge production does not erode either ecological or societal sustainability. This would be a truly sustainable agenda for new orientations of knowledge creation (as you might have noticed I prefer the term knowledge creation to production).

In more practical terms the challenge seems to be to enable concrete ways in which this can actually take place. The SuScit project can be seen as one attempt to open new democratic spaces for considering societal needs and what should be basic drivers of scientific development. But many others can be imagined, and are taking place, not least in terms of public engagement experiments on framing future research agendas on better meeting societal needs. To do so is a continuous practical challenge, and new orientations of science can only be developed through such practices, including concrete scientific disciplines and concrete societal settings and challenges. Clearly this cannot be reduced to a question of communicating science to the public. It is essentially a question of developing a double-orientation as part of scientific research and knowledge creation. This is why the approach of upstream public engagement in science is relevant in this respect: as an experiment not only on public engagement, but also in furthering new sustainable orientations of science.

Chapter 9
Findings and Calls for Action/Research

Throughout this book I have sought to explore how community-based action research in upstream public engagement can further new research orientations towards sustainability. I have been particularly interested in understanding how processes of social learning between citizens and scientists might further researchers in more reflexively taking into account complex social and environmental challenges faced in modern everyday life in deprived urban communities. To do so, I have been triangulating different theoretical perspectives in order to examine the empirical experiences of a particular case: the UK Citizen Science for Sustainability (SuScit) project. Addressing the overall question above, two particular research questions have been analysed throughout the book. First: How can empirical orientations towards sustainability in the SuScit process be understood and challenged by theoretical concepts of sustainability? Secondly: How can the SuScit process be understood in terms of social learning? In the following I shall combine the insights emerging from these analyses before reflecting on further perspectives for future action/research.

Difficulties of Science for Sustainability

In order to understand and challenge orientations towards sustainability in the SuScit process, it has been essential in this book to work critically with concepts of sustainability in the context of democracy. Doing so, I have approached the quest for sustainability as a challenge of overcoming what I have termed social environmental problems: historical challenges of our time, which are equally socially and environmentally grounded. In this sense, challenges of sustainability at societal level not least represent complex cultural questions concerning modern ways of living, and hence in democracies calling for democratic approaches to further sustainability. Science plays an ambiguous role in this context, not merely offering solutions to, but also making up an inherent part of modern socio-technological dynamics historically constituting present states of unsustainability. To challenge understandings of how science can further sustainability in a democratic context, I have been contrasting three different theoretical perspectives: First, Elling's considerations over sustainability building on Habermas' understanding of modern rationality; second, Shiva's critical perspective contrasting contemporary understandings of scientific and technological progress with a global outlook on sustainability issues; and third, the framework of Critical Utopian Action Research by Nielsen and Nielsen for understanding social learning processes based on shared collaboration and everyday life experience.

On this basis, and in particular drawing on Shiva's thoughts, sustainability can be conceptualised, in its most basic sense, as an immanent and emergent capability of ecological and social life to continuously renew itself without eroding its own foundation for existence. To put it simply, sustain-ability is an ability of life. Thus, in a societal and cultural perspective, basic human activities of everyday life, historically embedded in societal structures, interdependently determine and echo whether society is moving in a sustainable direction. In this sense sustainability is not invented, but supported (or eroded) by scientific and technological progress.

In modern, socially out-differentiated societies another layer of complexity is added to this perspective. When essential aspects of human-nature relations become managed through institutionalised systems of expertise, in order to continuously respond to ever-changing societal and ecological challenges, these need to be able to transcend pre-determined systemic rationales. Because the challenge of sustainability basically concerns the way humans live on Planet Earth, expertise and knowledge-based systems therefore must be capable of reflexively taking into account and respond to interdependent dynamics of social environmental problems, not merely as systemic problem-solving, but as part of the dynamics of modern democracies. This is why mutual learning processes are needed to further sustainability at societal level. It is in this particular perspective that I have sought to understand how orientations towards sustainability emerged, in the interface between citizens and scientists, in the upstream public engagement process of the SuScit project.

The SuScit project basically consisted of two parts. First, a community film-making project, where residents in a deprived urban neighbourhood in London explored and shared their perceptions of what it is like to live in their local area and aspirations for the future. Second, a program of shared workshop collaboration for sustainability practitioners and researchers to actively engaging, listen and learn from these perspective and respond by developing a community-led agenda for urban sustainability research. Essential for the first part of the project was that residents through the community projects, on the basis of their everyday life experience, started to articulate not only what appeared as unsustainable in modern urban life, but also aspirations of what a different or more desirable future might look like. In a social learning perspective these aspects of social imagination are particularly interesting since they may be potential starting points for transcending challenges of the present into future ways of living. Whilst it was on the one hand clear that social challenges were given higher priority than environmental ones, examples of the community work also showed examples of how nature relations are socially embedded in everyday life. Certainly this does not imply a sustainable human-nature relation, but it highlights important potentials for integrating environmental and social sustainability in modern urban living, which can be explored and developed further.

A key aspect of the second part of the project was that researchers through the workshop collaboration, on the basis of these identifications of unsustainability, responded to local community issues by translating these into different academic contexts, thereby exploring how research might be able to respond to community needs. Whilst this process seems to hold some potential for initiating research more responsive to local sustainability issues, it was also evident that research is often embedded in institutionalised practices, which does not fit very well with the idea of furthering sustainability through community engagement. An example of this tension appeared in the researcher reflections on the project highlighting a common concern about engaging with local communities without epistemologically treating them like research laboratories. This divergence between furthering research and more directly engaging local communities was a central theme in the project, both concerning the dynamics of public engagement and the orientations of sustainability research.

These findings offer a critical perspective on contemporary academic considerations on the emergence of new and more contextualised modes of knowledge production as conceptualised in Nowotny, Scott and Gibbons' mode-2 theory. A more critical examination of the diverging orientations of science would probably gain from taking into consideration the implications of science increasingly being oriented towards intellectual commodity production driving the knowledge economy, rather than aiming to sustain the sustain-ability of local communities.

Social Learning by the Doubling of Free Space

In continuation of the above perspectives on science and sustainability, I have addressed the question of how the SuScit process can be understood in terms of social learning, as a prerequisite for research to more adequately take into account and respond to social environmental problems faced by local communities. To analyse this question I build on a theoretical concept of social learning developed by Nielsen and Nielsen. In this perspective an essential feature of the bottom-up community-based approach of the SuScit project was that it enabled a social arena for community perspectives to be articulated, shared and explored. In an action research perspective, this feature of the project can adequately be understood as the creation of a free space: social arenas in everyday life where dominant external power structures are delimited, enabling everyday life perspectives to be articulated. The epistemological value of establishing free space in everyday-life-based action research is that, what is normally being suppressed or marginalised by contemporary societal power structures, in which everyday life is embedded, can potentially be articulated and shared. From other action research projects it is evident that free spaces call for marginalised and suppressed voices to be articulated and shared. In this sense the community work of the SuScit project not least provided community participants an opportunity to highlight unsustainable features of modern urban living and aspirations for more desirable futures.

This particular aspect of the project, however, not only proved valuable as part of the community work. It also served a second function by impacting on the researchers' engagement with the community enabling what can be conceptualised as a free space in an academic context. Thus academic rationalities and discourses were disturbed and traditional academic agendas – more or less successfully – suspended in pre-determining the process. In this sense the facilitation of the process did not merely provide a free space for the residents, but to some extent also for the researchers involved, because a different arena for deliberating on sustainability emerged. To understand this feature of the process I suggest the notion of creation and doubling of free space to conceptualise that what first served as a free space at community level had a second function as a free space in an academic context.

The impact of this is not least reflected in researchers' reflections on the project, highlighting a tension between academic and community orientations. Faced with this tension, the community engagement process seemed to call for researchers' reflections on barriers and potentials for community perspectives on urban sustainability to be taken into account in sustainability research. In this sense the doubling of free space is potentially a productive social space transcending systemic rationalities of science. This is an essential finding in terms of conceptualising how action research in upstream public engagement methodologically might be able to further new research orientations towards sustainability: by furthering a double-orientation of research, both being fixated on a particular subject of research, and being able openly to consider its own basic presumptions and orientations, not just in terms of intra-scientific considerations, but also through societal and democratic deliberations.

Calls for Further Action/Research

In the beginning of this book I wrote that if we are in a historical transformation towards new modes of science, we might just be in the beginning of grasping what that might imply. In a sustainability perspective it would be valuable to further the findings of this book in a dual sense: to consolidate the experiences of the project, and to explore how they might be integrated into processes of knowledge creation.

The consolidation could basically imply further experiments on doubling free space in the area of upstream public engagement. Whilst the strength of the research methodology of this book is the possibility of exploring new perspectives on sustainability, it would be relevant to undertake more systematic replication and methodological development.

The integration could basically imply finding ways by which the approach of this book can be part of sustainability research. As long as public engagement initiatives are merely organised as temporary projects highly separated from the daily practices of knowledge production, these initiatives might provide novel perspectives, but hardly impact on the way research is being done. In this sense, although the field of upstream public engagement is an interesting field of furthering sustainable orientations in science, its real potential is to further new kinds of knowledge creation.

It is in this particular context, the methodology and theoretical foundation of Critical Utopian Action Research offers valuable insights for working with local communities in furthering sustainability, not just as a scientific but also as a long-term democratic project. In my view the action research concepts of social imagination, the common third, and the creation and doubling of free space are all aspects of social learning whereby community-based action research in upstream public engagement offers some clear potentials for furthering new research orientations towards sustainability.

References

Adorno, T., 1998: *Aestetic Theory*. University of Minnesota Press.

Adorno, T. and Horkheimer, M., 1972: *Dialectic of Enlightenment*. Herder & Herder.

Alveson, M. and Skjöldberg, K., 2000: *Reflexive Methodology. New Vistas for Qualitative Research*. SAGE Publications.

Arrington, R. (ed.), 2001: *A Companion to the Philosophers*. Blackwell Companions to Philosophy.

Bateson, G., 1979: *Mind and Nature. A Necessary Unity*. Hampton Press.

Bawden, R., Guijt, I. and Woodhill, J., 2007: The critical role of civil society in fostering societal learning for a sustainable world. In Wals, A. (ed.): *Social Learning. Towards a Sustainable World*. Wageningen.

Beck, U., 1992: *Risk Society: Towards a New Modernity*. SAGE Publications.

Beck, U., Giddens, A. and Lash, S., 1994: *Reflexive Modernisation. Politics, Tradition and Aesthetics in the Modern Social Order*. Stanford University Press.

Becker-Smith, R., 1982: Modsætningsfyldt realitet og ambivalens: Kvinders arbejdserfaringer i fabrik og familie. Udkast. *Dansk tidsskrift for kritisk samfundsvidenskab*. Nr. 2. 1982. 10. Årgang, pp. 164–98.

Bidwell, D., 2009: Is Community-Based Participatory Research Postnormal Science? *Science, Technology and Human Values* 34(6) pp. 741–61.

Blackmore, C. (ed.), 2010: *Social Learning Systems and Communities of Practice*. Springer.

Boltanski, L. and Chiapello, E., 2005: *The New Spirit of Capitalism*. Verso.

Boutang, Y., 2007: *Le Capitalisme Cognitif: La Nouvelle Grande Transformation*. Multitudes/Idées. Editions Amsterdam.

Bruno, G., 1584: *De la causa, principio et uno*. Venice [e.g. Greenberg, S., 2011: *The Infinite Giordano Bruno*. Literary Licesing LCC].

Brydon-Miller, M., 2008: Ethics and Action Research: Deepening our Commitment to Principles of Social Justice and Redefining Systems of Democratic Practice. In Reason, P. and Bradburry, H.: *The SAGE Handbook of Action Research*. SAGE Publications.

Brydon-Miller, M., Antal, B., Friendman, V. and Wicks, P., 2011: The changing landscape of arts and action research. *Action Research* 9(1) pp. 3–11.

Buck, S., 1998: *The Global Commons*. Earthscan.

Chisholm, F. and Elden, M., 1993: Features of emerging Action Research. *Human Relations* 46(2) pp. 275–97.

Cooke, B. and Kothari, U. (eds), 2001: *Participation: The New Tyranny?* Zed Books.

Cruikshank, B., 1999: *The Will to Empower: Technologies of Citizenship and the War on Poverty*. Cornell University Press.

Drewes, L., 2006: The Methods and Implications of Action Research. A Comparative Approach to Search Conferences, Dialouge Conferences and Future Workshops. In Nielsen, K. and Svensson, L.: *Action Research and Interactive Research. Beyond Practice and Theory*. Shaker Publishing.

Dryzek, J., 1997: *The Politics of the Earth. Environmental Discourses*. Oxford University Press.

Dyball, R., Brown, V. and Keen, M., 2007: Towards sustainability: Five Strands of Social Learning. In Wals, A. (ed.): *Social Learning. Towards a Sustainable World*. Wageningen.

Eames, M., 2005: *Reconciling Environmental and Social Concerns*. Joseph Rowntree Foundation.

Eames, M. and Egmose, J., 2011: Community Foresight for Urban Sustainability: Insights from the Citizen Science for Sustainability (SuScit) Project. *Technological Forecasting and Social Change* 78(5) pp. 769–84.

Eames, M., Egmose, J., Adebowale, M. and Iudicissa, I., 2009a: *Towards a Community-led Agenda for Urban Sustainability Research: Insights from the Citizen Science for Sustainability (SuScit) Project*. Brunel University.

Eames, M., Egmose, J., Adebowale, M. and Iudicissa, I., 2009b: *Community Engagement for Science and Sustainability: Insights from the Citizen Science for Sustainability (SuScit) Project*. Brunel University.

Eikeland, O., 2006: The Validity of Action Research – Validity in Action Research. In Nielsen, K. and Svensson, L. (eds): *Action Research and Interactive Research. Beyond Practice and Theory*. Shaker Publishing.

Elling, B., 2010: A Record on Modernity, Rationality and Sustainability. In Nielsen, K., Elling, B., Figueroa, M. and Jelsøe, E. (eds): *A New Agenda for Sustainability*. Ashgate.

Elling, B., 2008: *Rationality and the Environment. Decision-making in Environmental Politics and Assessment*. Earthscan.

European Commission, 2013: Science for an Informed, Sustainable and Inclusive Knowledge Society. *Policy paper by President Barroso's Science and Technology Advisory Council*. Brussels, 29 August 2013.

European Commission, 2002: Science and Society Action Plan. Luxembourg: *Office for Official Publications of the European Communities*.

European Commission, 2001: European Governance: A White Paper. COM (2001) 428 FINAL, 25 July 2001. *Brussels: European Commission*.

Felt, U., Fochler, M., Müller, A. and Strassig, M., 2009: Unruly ethics: on the difficulties of a bottom-up approach to ethics in the field of genomics. *Public Understanding of Science* 18(3) pp. 354–71.

Fischer, F., 2000: *Citzens, Experts, and the Environment: The Politics of Local Knowledge*. Duke University Press.

Flyvbjerg, B., 1998: *Rationality and Power: Democracy in Practice*. The University of Chicago Press.

Funtowicz, S. and Ravetz, J., 1999: Post-normal Science – an insight now maturing. *Futures* 31 pp. 641–6.

Funtowicz, S. and Ravetz, J., 1993: Science for the post-normal age. *Futures* 25(7) pp. 739–55.

Gibbons, M., Limoges, C., Nowotny, H., et al., 1994: *The New Production of Knowledge – the Dynamics of Science and Research in Contemporary Societies*. Sage Publications.

Greenwood, D., 2009: Are Research Universities Knowledge-Intensive Learning Organisations? In Jemielniak, D. and Kociatkiewicz, J.: *Handbook of Research on Knowledge-Intensive Organisations*. Information Science Reference.

Greenwood, D. and Levin M., 2007: *Introduction to Action Research: Social Research for Social Change*, second edition revised. Sage Publications, Inc.

Habendijk, R. and Irwin, A., 2006: Public deliberation and governance: Engaging with science and technology in contemporary Europe. *Minerva* 44 pp. 167–84.

Habermas, J., 1981: *Theory of Communicative Action*. Beacon Press.

Habermas, J. 1968: *Technik und Wissenschaeft als 'ideologie'*. Surhkamp Verlag.

Hawken, P., Lovins, A. and Lovins, L., 1999. *Natural Capitalism. The Next Industrial Revolution*. Earthscan.

Høyrup, J., 2002: *From Hesoid to Saussure, From Hippocrates to Jevons. An Introduction to the History of Scientific Thought*. Roskilde University. Section for Philosophy and Science Studies.

Irwin, A., 2006: The politics of talk: Coming to Terms with the 'New' Scientific Governance. *Social Studies of Science* 36(2) pp. 299–320.

Irwin, A., 2001: Constructing the Scientific Citizen: Science and Democracy in the Biosciences. *Public Understanding of Science* 10(1) pp. 1–18.

Islington Council, 2006: *The One Islington Corporate Plan 2006–2009*.

Islington Strategic Partnership, 2006a: *Islington's Local Area Agreement 2006 to 2009*.

Jackson, T., 2009: *Prosperity Without Growth?* Sustainable Development Commission.

Jungk, R. and Müllerts, R., 1987: *Future Workshops. How to Create Desirable Futures*. Institute for Social Inventions.

Kamara, M., 2009: The Typology of the Game that American, British, and Danish Scientists Play. *Minerva* 47(4).

Kirkeby, Per 1996: *Vinterbillede*. Jesper Jargil Film. Statens Filmcentral.

Klüver, L., Nentwich, M., Peissl, W., et al., 2000: *European Participatory Technology Assessment. Participatory Methods in Technology Assessment and Technology Decision-Making*. Europta. The Danish Board of Technology.

Kristensen, J., 2008: Kapitalismens nye ånd og økonomiske hamskifte. *Dansk Sociologi* 19(2).

Latour, B., 1998: From the World of Science to the World of Research? *Science Magazine* 280(5361) pp. 208–9.

Lucas, K., Fuller, S., Psaila, A. and Thrush, D., 2004: *Prioritising Local Environmental Concerns: Where There's a Will There's a Way*. Joseph Rowntree Foundation/ York Publishing Ltd.

Lucas, K., Walker, G., Eames, M., et al., 2004: *Environment and Social Justice: Rapid Research and Evidence Review*. Sustainable Development Research Network, PSI.

Meadows, D., Meadows, D. and Randers, J., 1972: *The Limits to Growth*. Universe Books.

Mills, W., 1959: *The Sociological Imagination*. Oxford University Press.

Moll, A., Sonnenfeld, D. and Spaargaren, G. (eds), 2009: *The Ecological Modernisation Reader. Environmental Reform in Theory and Practice*. Routledge.

Morell, L., 2004: *Kunstneren som polyhistor – den intellektuelle overbygning i Per Kirkebys værk*. Århus Universitetsforlag.

Negt, O., 1985: *Det levende arbejde, den stjålne tid*. Politisk Revy, København.

Negt, O., 1975: *Sociologisk fantasi og eksemplarisk læring* [*Soziologishe Phantasie und exemplarishes Lernen*]. Roskilde Universitstsforlag.

Negt, O. and Kluge, A., 1972 [1974]: Offentlighed og erfaring. Surhkamp Verlag. *Nordisk sommeruniversitets skriftserie* 3.

Nielsen, B., 2010: Sansebevidsthed og gestisk viden. Om det æstetiske ud fra kritisk teori. *Dansk Pædagogisk Tidsskrift* 2 pp. 24–33.

Nielsen, B. and Nielsen, K., 2015: Critical Utopian Action Research. The Potentials of Action Research in Democratization of Society. In Gunnarsson, E., Hansen, H. and Nielsen, B. (eds): *Action Research for Democracy*. [*Under submission*]

Nielsen, B. and Nielsen, K., 2007: *Demokrati og naturbeskyttelse. Dannelse af borgerfællesskaber gennem social læring – med Møn som eksempel*. Frydenlund.

Nielsen, B. and Nielsen, K., 2006a: Methodologies in Action Research. In Nielsen, K. and Svensson, L.: *Action Research and Interactive Research. Beyond Practice and Theory*. Shaker Publishing.

Nielsen, B. and Nielsen, K., 2006b: *En menneskelig natur. Aktionsforskning for bæredygtighed og politisk kultur*. Frydenlund.

Nielsen, B., Nielsen, K. and Olsén, P., 1999: *Demokrati som læreproces. Industri og lykke: Et år med Dyndspringeren*. Roskilde Universitets Forlag.

Nielsen, B., Nielsen, K. and Olsén, P., 1996: From Silent to Talkative Participants: A Discussion of Technique as Social Construction. *Economic and Industrial Democracy* 17 pp. 359–86.

Nielsen, K., 2001: At frisætte subjektivitet. Om sociale eksperimenters betydning som kvalitativ forskningsmetode i humanvidenskaberne. *Nordiske udkast* 1.

Nielsen, K., 1997: Ekspertkultur og orienteringsusikkerhed. In Weber, K., Nielsen, B.S. and Olesen, H.S. (eds): *Modet til fremtiden – inspirationen fra Oskar Negt*. Roskilde Universitstsforlag.

Nielsen, K., 1996: *Arbejdets sociale orientering. En undersøgelse af forholdet imellem arbejdslivsdemokrati og mulighederne for bæredygtighed i industrielle moderniseringsprocesser.* Forlaget Sociologi.

Nielsen, K., Elling, B., Figueroa, M. and Jelsøe, E. (eds), 2010: *A New Agenda for Sustainability.* Ashgate.

Nielsen, K. and Svensson, L., 2006: *Action Research and Interactive Research. Beyond Practice and Theory.* Shaker Publishing.

Nowotny, H., Peter, S. and Gibbons, M., 2001: *Rethinking Science: Knowledge in an Age of Uncertainty.* Polity.

Ostrom, E., 1990: *Governing the Commons: The Evolution of Institutions for Collective Action.* Cambridge University Press.

Ostrom, E. and Hess, C. (eds), 2006: *Understanding Knowledge as a Commons: From Theory to Practice.* MIT Press.

Postmand, N., 1992: *Teknopolis. Kulturens knæfald for teknologien.* Gyldendal.

Pålshaugen, Ø., 1998: *The End of Organization Theory.* John Benjamins Publishing.

Ramage, M. and Shipp, K., 2009: *Systems Thinkers.* Springer Publishing.

Reason, P. and Bradbury, H., 2008: *Handbook of Action Research. Participative Inquire and Practice.* SAGE Publications.

Sachs, W., 2010: *Environment.* In Sachs, W. (ed.): *The Development Dictionary.* Zed Books.

Shinn, T., 2002: The Triple Helix and New Production of Knowledge: Prepackaged Thinking on science and Technology. *Social Studies of Science* 32(4) pp. 599–614.

Shiva, V., 2005: *Earth Democracy. Justice, Sustainability and Peace.* South End Press.

Skjervheim, H., 1996 (1957): *Participant and Spectator.* Skriftserien nr. 12. The department of philosophy. Bergen.

Smith, G., 2009: *Democratic Innovations. Designing Institutions for Citizen Participation.* Cambridge University Press.

Stilgoe, J., 2009: *Citizen Scientists. Reconnecting Science with Civil Society.* DEMOS.

Tofteng, D. and Husted, M., 2011: Theatre and action research: How drama can empower action research processes in the field of unemployment. *Action Research* 9(1) pp. 27–41.

Tofteng, D. and Husted, M., 2006: The common third. In Nielsen, K. and Svensson, L.: *Action Research and Interactive Research. Beyond Practice and Theory.* Shaker Publishing.

Wals, A. (ed.), 2007: *Social Learning. Towards a Sustainable World.* Wageningen Academic Publishers.

Weber, K., Nielsen, B. and Olesen, H. (eds), 1997: *Modet til fremtiden – inspirationen fra Oskar Negt.* Roskilde Universitstsforlag.

Westlander, G., 2006: Researcher roles in action research. In Nielsen, K. and Svensson, L.: *Action Research and Interactive Research. Beyond Practice and Theory.* Shaker Publishing.

Wilsdon, J. and Willis, R., 2004: *See-through Science. Why Public Engagement Needs to Move Upstream.* Demos.

Wilsdon, J., Wynne, B. and Stilgoe, J., 2005: *The Public Value of Science. Or How to Ensure that Science Really Matters.* Demos.

World Commission on Environment and Development, 1987: *Our Common Future.* Oxford University Press.

Wynne, B., 2006: Public Engagement as a means of Restoring Public Trust in Science – Hitting the Notes, but Missing the Music? *Community Genet* 327-T2. Karger. Basel.

Wynne, B., 2005: Risk as globalizing 'democratic' discourse? Framing subjects and citizens. In Leach, M., Scoones, I. and Wynne, B.: *Science and Citizens. Globalization & the Challenge of Engagement.* Zed Books.

Empirical Research

Empirical research materials from the Citizen Science for Sustainability (SuScit) project and subsequent participant interviews and are indicated by the following codes:

Project phases

p01- Desktop research and state of the art
p01- Preparation of fieldwork programme
p02- Establishing partnerships and recruiting participants
p03- Participant introduction meetings and focus groups
p04- Community filmmaking projects
p05- First shared workshop
p06- Second shared workshop
p07- Researcher and practitioner workshop
p08- Third shared workshop
p09- Mapping emerging projects and initiatives
p10- Project dissemination
p11- Project analysis and reports
p12- Subsequent participant interviews

Participant groups

YP The Young Peoples' Panel
LP The Lone Parents Panel
OP The Older Peoples' Panel

R The Researchers' Panel
P The Practitioners' Panel

Individual participants

Individual participants are numbered subsequent to the participant group.

Example: (Empirical research p12-R2) refers to the evaluation interview (project phase p12) with the second member (individual 2) of the Researchers Panel (group R).

Index

Please note: References to figures are listed in *italics.*

For Product Safety Concerns and Information please contact our EU
representative GPSR@taylorandfrancis.com
Taylor & Francis Verlag GmbH, Kaufingerstraße 24, 80331 München, Germany

www.ingramcontent.com/pod-product-compliance
Ingram Content Group UK Ltd.
Pitfield, Milton Keynes, MK11 3LW, UK
UKHW011019180926
475701UK00004B/828
* 9 7 8 0 3 6 7 3 3 2 8 8 4 *